SOU

ALL ABOUT BIOLOGY

Brian Beckett &
RoseMarie Gallagher

Oxford University Press

INTRODUCTION

Biology is all about living things . . . what they do, and how, and why. Today's biologists must also tackle important topics like pollution, the environment, population, and health. This book will help you to answer some of the questions of biology and encourage you to ask more.

How to use this book

Everything in this book has been organized to help you find things quickly and easily. It is written in two-page units. Each unit is about a topic that you are likely to study.

- Use the contents page

If you are looking for information on a large topic, look it up in the contents list. But if you cannot see the topic you want then:

- Use the index

If there is something small you want to check on, look up the most likely word in the index. The index gives the page number where you'll find more information about that word.

- Test yourself

There are questions at the end of each topic. Working through these will help you to check that you have read and understood the topic properly.

- Investigate further

At the end of each chapter there are investigations for you to try. These will help you develop the practical skills a biologist needs.

We hope that when you complete this book you will have a greater understanding of what biology is all about, and will have enjoyed studying it.

Brian Beckett
RoseMarie Gallagher

Oxford University Press, Great Clarendon Street, Oxford OX2 6DP
Oxford New York
Athens Auckland Bangkok Bogota Bombay
Buenos Aires Calcutta Cape Town Dar es Salaam
Delhi Florence Hong Kong Istanbul Karachi
Kuala Lumpur Madras Madrid Melbourne
Mexico City Nairobi Paris Singapore
Taipei Tokyo Toronto Warsaw
and associated companies in
Berlin Ibadan

Oxford is a trade marks of Oxford University Press

First published 1988
Reprinted 1990, 1992, 1993, 1994, 1995, 1996, 1998
ISBN 0 19 914070 7

Typeset by Best-set Typesetter Ltd, H.K.
Printed in Hong Kong

Contents

1·1 Living things

You are a **living thing**. Grass, whales, and bats are living things too. But stones and rain are non-living things. Living things are different from non-living things in the ways shown below.

Living things move and have senses

Animals walk, or run, or hop, or crawl, or swim, or fly. They find their way using **sense organs**. These are eyes, ears, noses, taste buds, skin, and insect feelers called **antennae**.

Plants move by growing, like these beans growing up bean poles. They don't have sense organs but they can still respond to things. Roots grow down in response to gravity, and to find water. Shoots grow up towards light.

Living things feed

They need food for energy, growth, and repair.

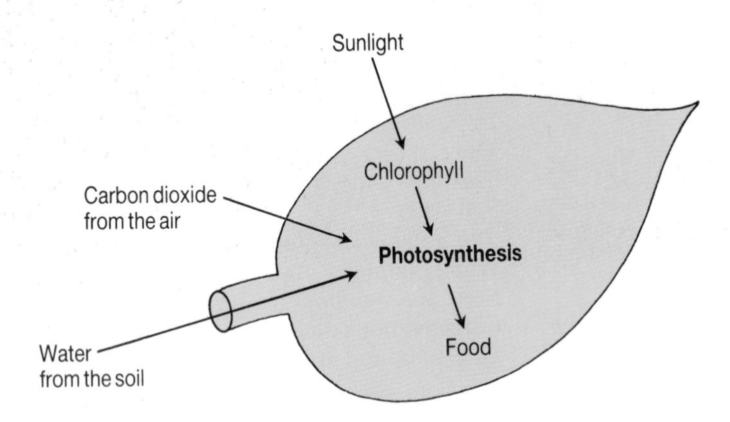

Plants make their own food in their leaves. This is called **photosynthesis**. It needs light, water, carbon dioxide, and a green chemical called **chlorophyll** which is found in leaves.

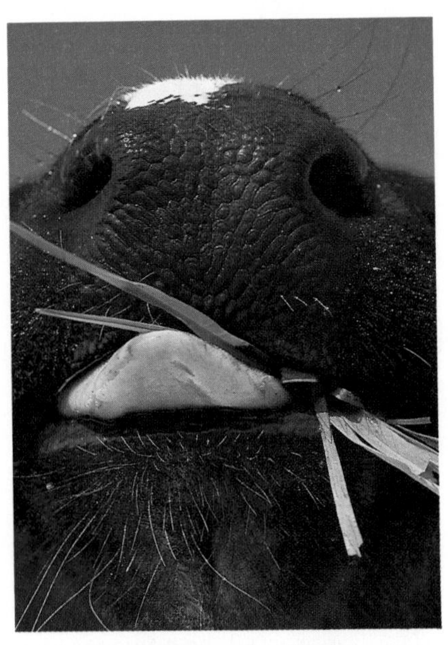

Animals can't make their own food so they eat plants and other animals. Which animal is this? What is it eating?

Living things respire

They get energy from food by a process called **respiration**. This usually needs oxygen.

Food + Oxygen ⇨ ENERGY + Waste: water and carbon dioxide

Living things excrete

All living things produce waste. The removal of waste from their bodies is called **excretion**.

Animals excrete through their lungs and kidneys, and through their skin when they sweat.

Plants store waste in old leaves, which fall in the autumn.

Living things reproduce and grow

Animals lay eggs, or have babies. Seeds from plants grow into new plants.

Animals stop growing when they reach their adult size.

Plants grow all their lives. This giant redwood tree has been growing for over 2000 years!

Questions

1 Name seven ways in which living things are different from non-living things.

2 Name the green stuff, and three other things plants need to make food.

3 Name all your sense organs.

4 What is:
   a) respiration?
   b) excretion?

1·2 Sorting and naming

Spiny anteater.

Jumping spider.

These are just two of the things that live on Earth. Altogether there are over one-and-a-half million different kinds of living thing!

To make it easier to study them biologists sort living things into groups which are alike in some way. Try sorting the six creatures below into groups. Do this now.

You could sort them into those found on land, in the air, and in water. But this would group together very different creatures like crabs, sea weeds, and fish. Would it make sense to sort them according to colour and size? What other ways are there of sorting them?

Biologists first sort living things into three large groups. The three large groups are:

1 **Simple organisms**	2 **Plants**	3 **Animals**

These huge groups are then split into smaller groups. You can see how this is done in the rest of this chapter.

Naming living things

If you want to find the name of a plant or animal you could look through books until you see a picture of it. Or you could use a **key**.

Keys

The simplest keys are made up of short, numbered sentences arranged in pairs. Look at the example below. Read the instructions, then use this key to name the insects drawn on this page.

How to use the key

Read the first pair of descriptions and decide which fit the insect you are trying to name. Opposite the description you choose there is a number. This number tells you which pair of descriptions to read next. Read them, and again decide which describes the insect. Opposite, you will find either the insect's name, or the number of the next pair of descriptions to read. Carry on until you find the insect's name.

An example of a key

1	Wings visible	3
	Wings not visible	2
2	Three-pronged tail	Bristle tail
	Pincers at end of tail	Earwig
3	Two pairs of wings	4
	One pair of wings	5
4	Wings fringed with hairs	Thrip
	Wings not fringed with hairs	6
5	Legs longer than body	Cranefly
	Legs not longer than body	Housefly
6	Wings larger than body	Butterfly
	Wings not larger than body	Wasp

1·3 Groups of living things

In this unit you can see how living things are first sorted into huge groups, which are then split into smaller and smaller groups.

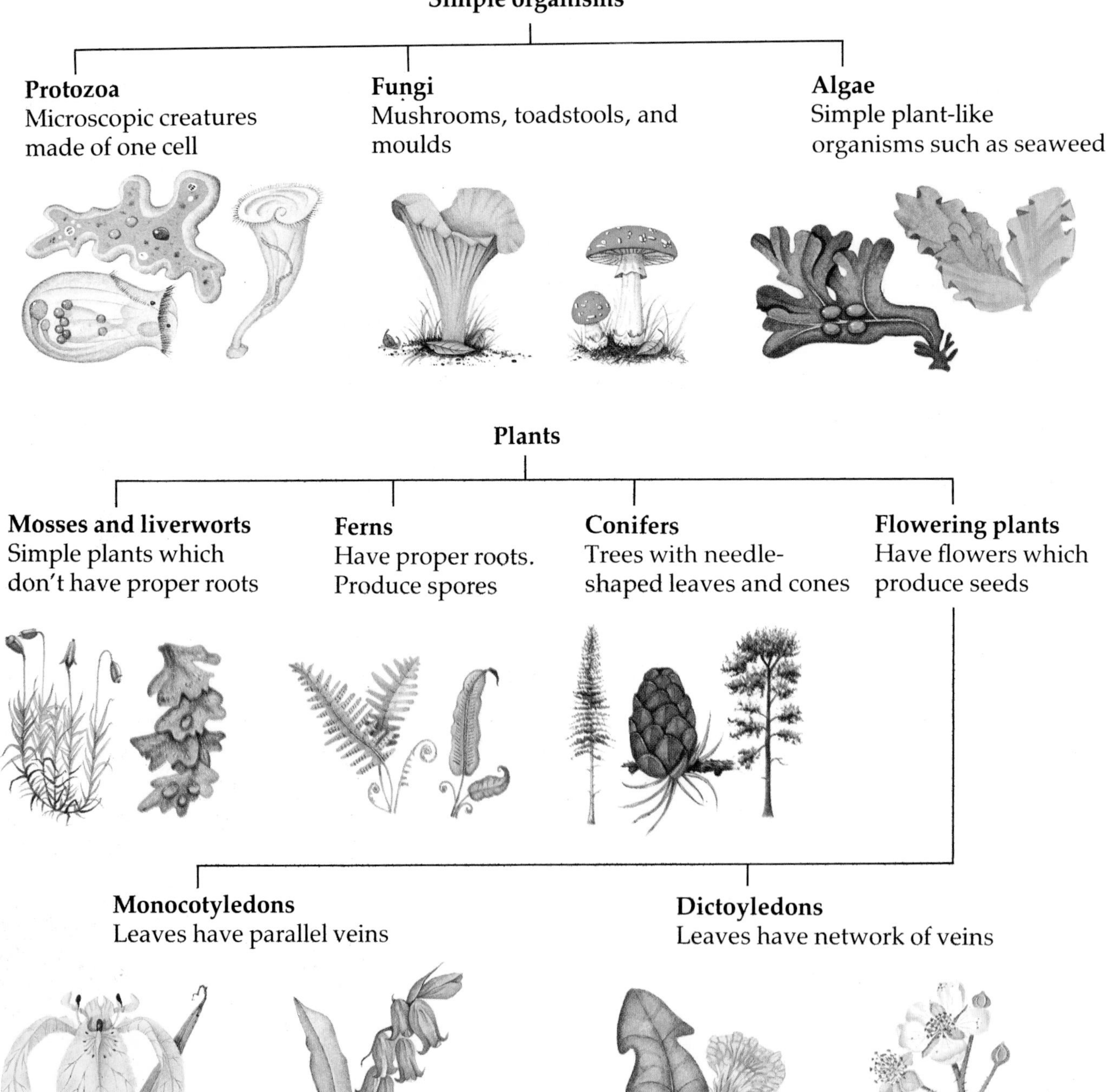

Animals without backbones (Invertebrates)

Coelenterates
Sack-like body with tentacles

Flatworms
Flat body covered with cilia

Annelid worms
Body encircled with many rings

Molluscs
One coiled shell or two uncoiled shells

Arthropods

Crustaceans
Hard outer skeleton

Insects
Six legs and usually two pairs of wings

Arachnids
Eight legs. No wings

Myriapods
Many pairs of legs

Animals with backbones (Vertebrates)

Fish
Scales, fins, and gills

Amphibia
live in water and on land

Reptiles
Covered with hard, dry scales

Birds
Covered with feathers. Wings

Mammals
Hairy skin. Young feed on mother's milk

Questions

1. What is the word for:
   a) animals with backbones?
   b) animals without backbones?
2. Do fungi belong to the plant group?
3. Which group do you belong to?
4. Which animals have four pairs of legs?
5. Which plants don't have flowers?
6. What is a dicotyledon?

1·4 Simple organisms

Protozoa

Protozoa are tiny creatures. Some are so small you could get thousands of them on the head of a pin. You need a microscope to see them properly. Some protozoa feed on other tiny creatures. Others contain chlorophyll, and can make their own food, like plants do. *Amoeba*, *Paramecium*, and *Euglena* are protozoa.

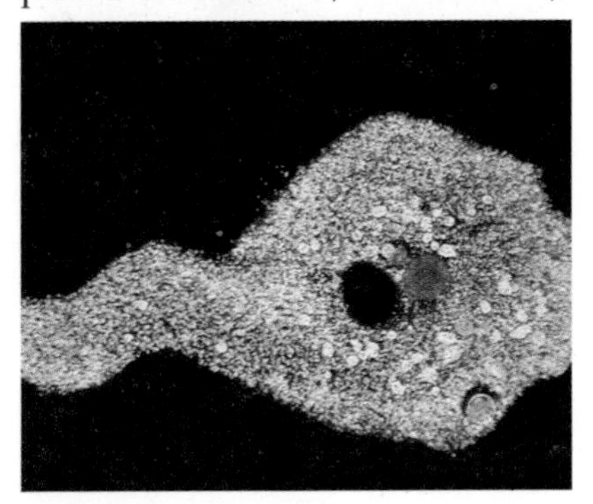

This is an *Amoeba*. It moves by changing its shape. It feeds on bacteria and other tiny creatures.

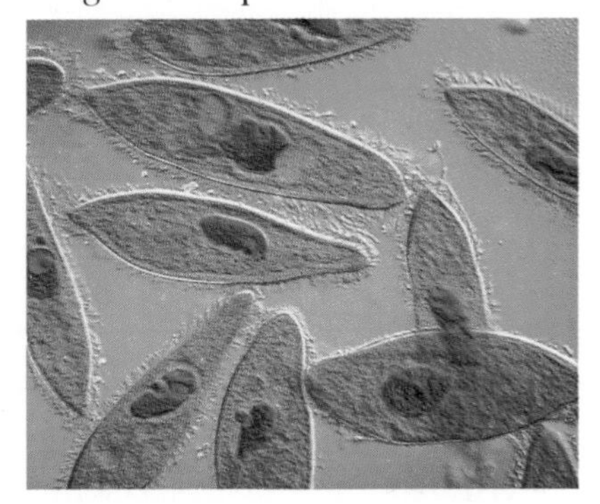

Paramecium has tiny hairs called cilia all over it. These trap its food and help it move.

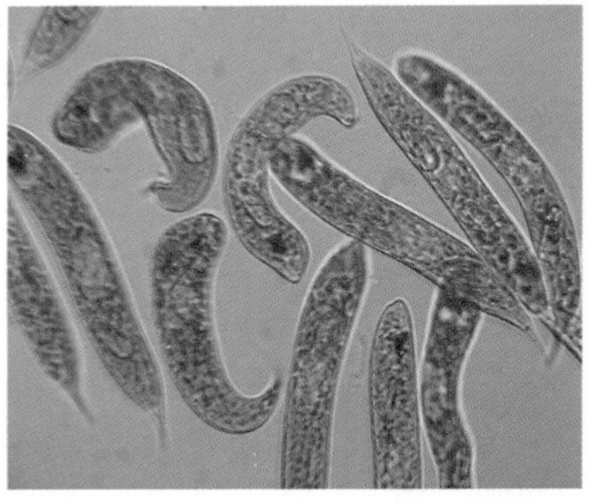

Euglena can make its own food, like plants do, because it contains chlorophyll. That's what makes it green.

Fungi

Mushrooms, toadstools, mildews, and moulds are all **fungi**. Some fungi look a bit like plants, but they cannot make their own food. Some fungi feed on dead things, like the remains of plants and animals. They are called **saprophytes**. Others feed on living things, and cause disease. They are called **parasites**. Ringworm in humans and mildew in plants are caused by parasitic fungi.

These mushrooms are fungi you can eat. Mushrooms do not need light, so some people grow them in cellars for sale to shops.

This is a poisonous fungus called a Fly Agaric. It is common in pine and birch woods. If eaten in large amounts it can kill you.

Bracket fungi kill trees, then digest the dead wood. You usually see them on dead trees.

All fungi are made up of fine threads called **hyphae** (pronounced hi-fee). It is easy to see the hyphae in moulds, but in mushrooms and toadstools hyphae are packed tightly together.

Fungi reproduce by forming **spores.** These are tiny cells that get scattered by the wind. When they land in the right place they grow into more fungi.

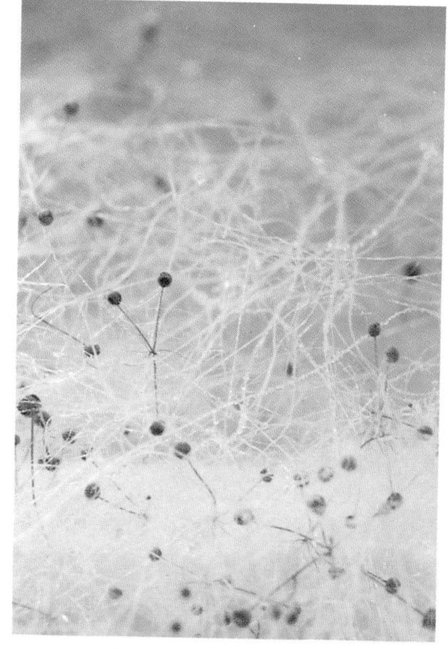

Bread mould growing on a piece of bread. The white threads are the hyphae. Each black dot contains thousands of spores.

These are puff-ball fungi. When a drop of rain hits the puff-ball, a cloud of spores shoots out.

Algae

Algae are like plants because they make their own food by photosynthesis. But they are not really plants because they do not have proper roots, stems, or leaves.

Seaweeds are large algae that grow in the sea. They can be brown, green, or red. This type is called bladderwrack.

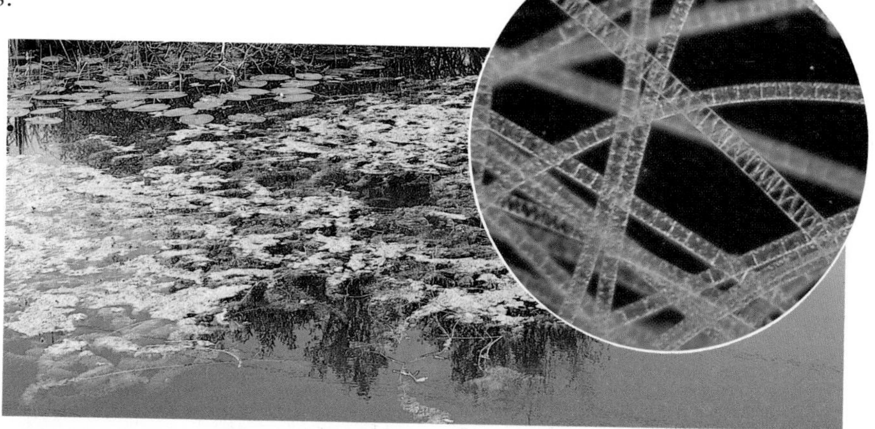

The green scum you sometimes see floating on ponds is made of tiny algae called Spirogyra. If you look at Spirogyra under a microscope you will see lots of tiny green threads. They are covered with slime which stops them getting tangled.

Questions

1 What are protozoa?
2 Name a protozoan that can make its own food.
3 Some fungi are saprophytes. What does this mean?
4 Are mushrooms saprophytes or parasites?
5 How do fungi reproduce?
6 In what way are algae like plants?
7 In what way are algae different from plants?
8 Name two kinds of algae. Where are these algae found?

1·5 Look no backbones! I

In this unit and the next, we look at groups of animals without backbones that you met on page 7. Animals without backbones are called **invertebrates**.

Coelenterates

Hydra, jellyfish, and sea anemones are all **coelenterates**. A coelenterate has a body like a bag. The open end is the mouth. It has tentacles round it to catch small animals for food. The tentacles have sting cells to sting the animals and paralyse them.

Hydra lives in ponds. It moves by turning cartwheels. It grows buds which break off to form new hydra.

This is a sea anemone. It uses sting cells on its tentacles to paralyse small creatures which are then pushed into its mouth.

Jellyfish live in the sea. They move by opening and closing like an umbrella. Some have sting cells which can hurt swimmers.

Flatworms

Flatworms have flat bodies with a mouth at one end. Freshwater flatworms, tapeworms, and flukes all belong to this group.

Fresh water flatworms live in ponds and streams. They move by waving tiny hairs called cilia.

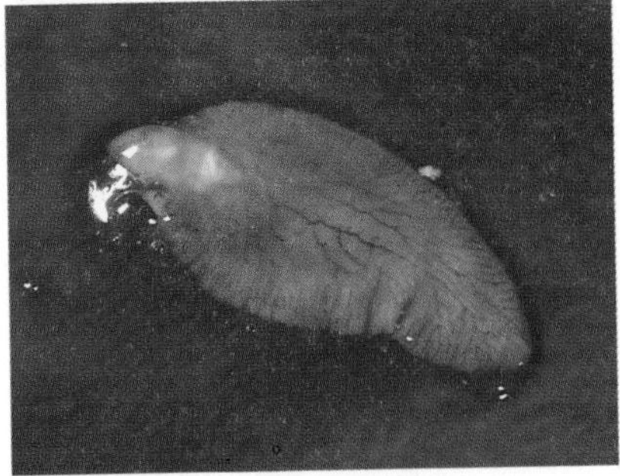

Flukes live inside other animals, and can cause disease. This shows a liver fluke on a sheep's liver. It hangs on by suckers. It can kill the sheep.

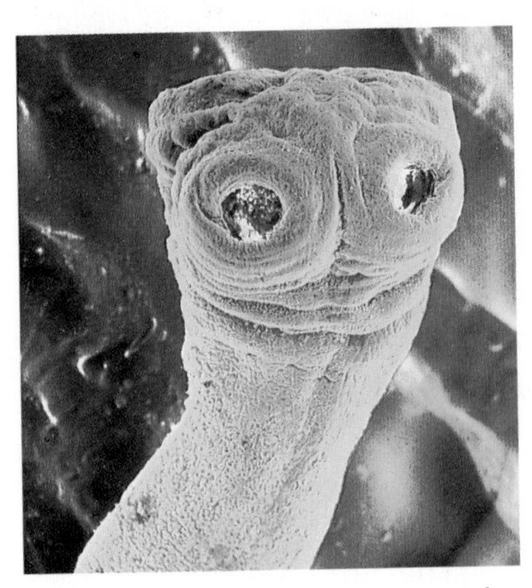

Tapeworms live in the guts of other animals, and eat their digested food. This one lives in humans. It can grow up to eight metres long.

Annelid worms

These have bodies made of rings, or **segments**. They are sometimes called true worms. Examples are earthworms, lugworms, and leeches.

Earthworms 'eat' soil and digest tiny creatures and dead leaves in it. The rest passes through their bodies.

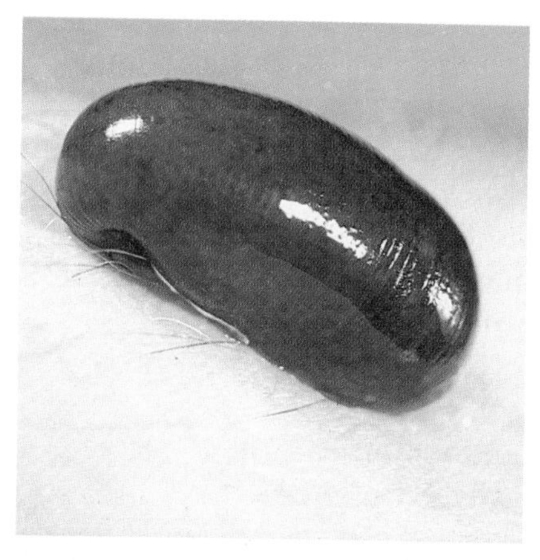

Leeches live on other animals and suck their blood. One type is used by surgeons to clean up infected flesh round wounds.

Lugworms live under sand on the seashore. They suck in sea water and eat any tiny creatures in it.

Molluscs

Snails, mussels, and octopuses are all molluscs. Molluscs have soft bodies. Most have one or two shells.

Snails have coiled shells. They move about on a slimy flat foot. They have long rough tongues for scraping up food.

A mussel has two shells hinged together. It anchors itself to rock by tough threads. It feeds on tiny creatures in sea water.

An octopus has a shell inside its body. It uses tentacles with suckers to move and to catch food.

Questions

1 Which group of animals has a body like a bag?
2 How do coelenterates feed themselves?
3 What do tapeworms eat?
4 What is the difference between flatworms and annelid worms?
5 Give another name for annelid worms.
6 Name three kinds of annelid worm.
7 Write down two features of mollucsc. Name two molluscs.
8 How do snails eat?

1·6 Look no backbones! II

Let's look now at the largest group of animals without backbones – the **arthropods**. Spiders, flies, and shrimps are all arthropods. Arthropods have a tough skin called a **cuticle**. They have jointed legs, and most have feelers or **antennae**. They can have simple eyes with only one lens, or **compound eyes** with thousands of lenses. They are divided into four smaller groups.

Like all flies, the horsefly is an arthropod. Look at its eyes. They are compound eyes, made up of thousands of tiny lenses.

Crustaceans

Crabs, lobsters, shrimps, woodlice, and waterfleas are **crustaceans**. Crustaceans have two pairs of antennae.

Crabs have a thick, hard cuticle and five pairs of legs.

Woodlice have a thin cuticle and live in damp places.

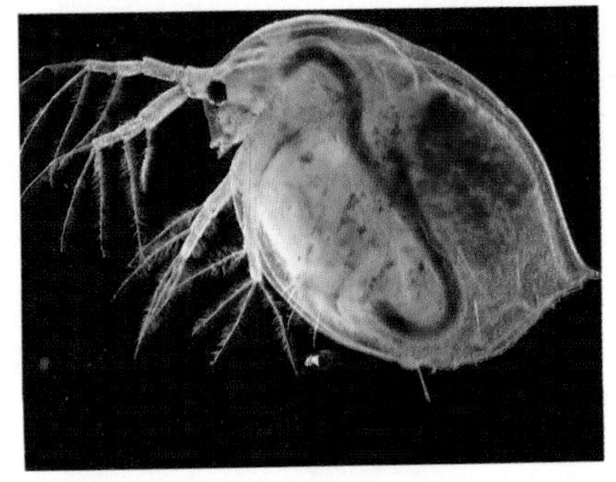

In waterfleas the cuticle is quite thin and soft.

Insects

For every person alive there are about a million insects! About 700 000 different kinds are known. Insects' bodies have three parts: a **head**, a **thorax**, and an **abdomen**.

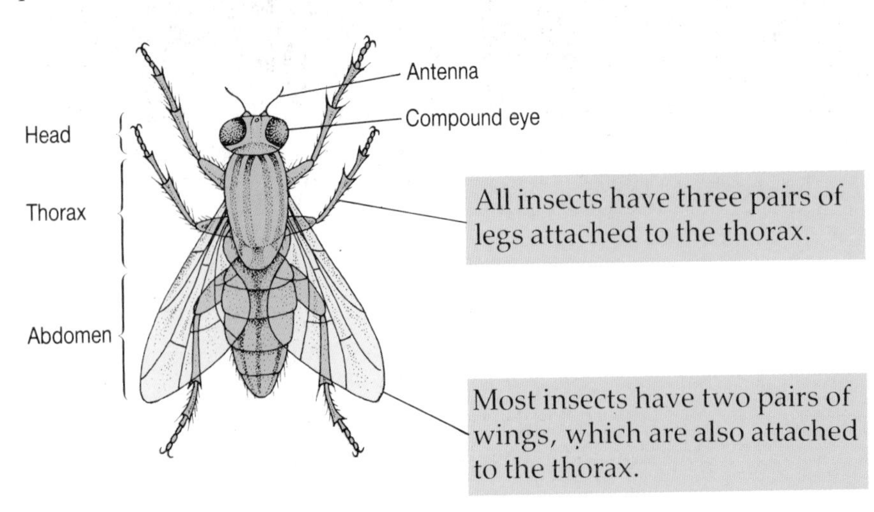

All insects have three pairs of legs attached to the thorax.

Most insects have two pairs of wings, which are also attached to the thorax.

Name as many parts of this horsefly as you can, by comparing it with the diagram opposite.

Many insects cause harm. Some destroy crops. Some spread disease. For example mosquitos spread malaria among humans. Some are a nuisance, like the midges that bite you in summer, and moths that damage clothes.

Many insects undergo big changes during their lives. Every butterfly was once a caterpillar. Every housefly was a maggot!

Colorado beetles are feared by potato growers because they destroy potato crops.

Many butterflies are brightly coloured. Scientists think they might use their colours to signal to each other.

Arachnids

Arachnids are arthropods with four pairs of legs, and no antennae. They use poison fangs to paralyse their prey. Spiders, harvestmen, mites, ticks, and scorpions are all arachnids.

Spiders are the most common arachnids. Many of them spin webs or tripwires of silk, to catch their prey.

A scorpion ready to strike. Its sting is in its tail. The sting from some scorpions can kill humans.

Myriapods

The word 'myriapod' means 'many pairs of feet'. Myriapods have long bodies made up of many segments. Each segment has at least one pair of legs. Centipedes and millipedes are myriapods.

Centipedes have one pair of legs to each segment. They paralyse their prey with poison fangs.

Millipedes have two pairs of legs to each segment. They eat plants.

Questions

1 Which group of arthropods has:
   a) three pairs of legs?
   b) four pairs of legs?
   c) two pairs of antennae?

2 Give two reasons why spiders are not insects.

3 Name two kinds of myriapod.

4 Name the three parts of an insect's body.

1·7 Animals with backbones I

In this unit and the next, we look at the groups of animals with backbones. These are called **vertebrates**. There are five groups of vertebrates:

fish	**amphibians**	**reptiles**	**birds**	**mammals**

You belong to the last group!

Fish

The body of a fish is smooth and streamlined for easy movement through water. Fish are covered with scales, like the tiles on a roof.

Most fish are weightless in water, because their bodies contain a bladder like a balloon which buoys them up.

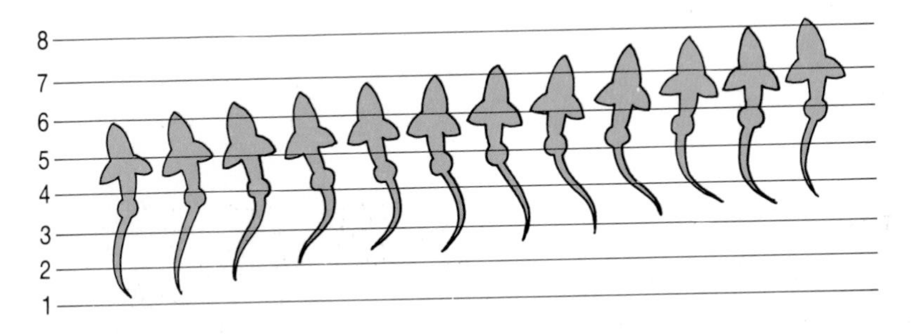

A fish swims by moving its tail from side to side.

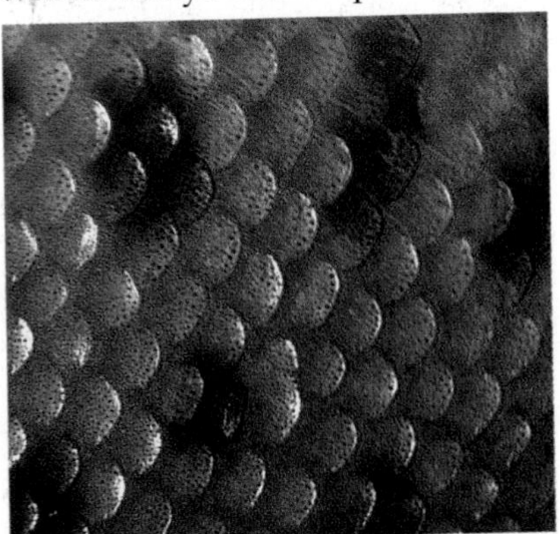
The scales on a trout's body.

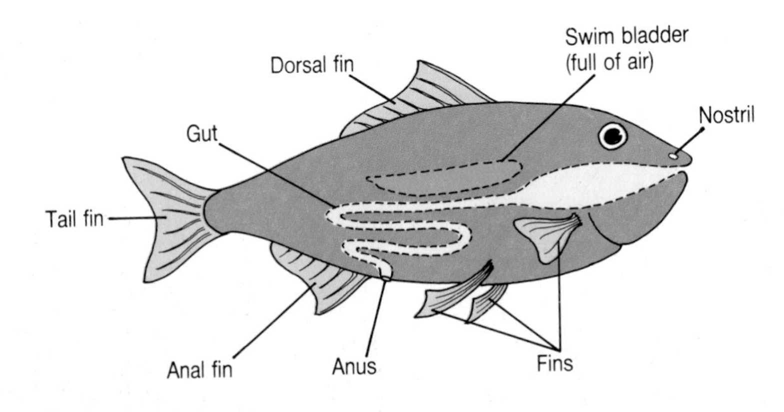

The parts of a fish.

Most female fish, like this trout, reproduce by laying eggs.

But these pregnant guppies will give birth to live baby fish.

A baby dogfish is born in a capsule with a sac of food.

Amphibians

Frogs, toads, salamanders, and newts are all **amphibians**.

Amphibians have four limbs and moist skin. They can live both on land and in water. On land they breathe using lungs. In water they breathe through their skin.

Amphibians lay their eggs in water. These hatch into tadpoles, which swim using tails, and breathe through gills.

This frog has just laid her eggs as frog spawn. In about a week they'll become tadpoles.

Newts live on land for part of the year. But in spring they move into the water to breed.

Reptiles

Lizards, crocodiles, snakes, and tortoises are reptiles. Reptiles have tough, dry, scaly skin. They breathe with lungs. They lay eggs with tough shells like leather.

A snake egg hatches . . . and out comes the snake's head.

Crocodiles look peaceful and lazy. But don't be fooled . . .

Some snakes won't harm you. But a viper like this one can kill.

Questions

1 Fish are streamlined. How does this help them?

2 How do fish breathe?

3 How do fish swim?

4 How is a tadpole different from a frog?

5 Explain how frogs can live on land and in water.

6 Do all reptiles have legs?

7 How are reptile eggs different from bird eggs?

8 Name two amphibians and two reptiles.

1·8 Animals with backbones II

Like all birds, penguins have wings – but they cannot fly.

Let's now look at the last two groups of animals with backbones – birds and mammals. Birds and mammals are **warm blooded**. This means their bodies always stay around the same temperature, no matter what the weather is like. All other animals are **cold blooded**. Their bodies get hot and cold with the weather.

Birds

All birds have feathers on their bodies. They also have wings, but not all birds can fly. They have beaks for pecking and tearing food. They lay eggs covered with a hard shell.

Birds have light hollow bones, to help them fly.

Water birds live on fish and other water creatures . . .

. . . but this tawny owl is about to dine on a mouse.

Mammals

Mammals have hair on their bodies. Female mammals have breasts, or **mammary glands**, from which their young ones suck milk. There are three kinds of mammals.

Mammals that lay eggs

Mammals that lay eggs are called **monotremes**. The spiny ant-eater, duck-billed platypus, and opossum are monotremes.

When monotreme eggs hatch the babies are not fully formed. They are fed on milk for several weeks while their missing parts grow.

The duck-billed platypus lays its eggs in an underground nest.

Mammals with pouches

Some mammals carry their young in a pouch. They are called **marsupials**. Kangaroos, wallabies, and koala bears are marsupials.

Baby marsupials start growing in their mother's womb. Before they are fully formed they crawl out of the womb and into her pouch. The pouch has a teat in it from which the young suck milk.

A kangaroo carrying her baby in her pouch.

Placental mammals

Humans, bats, whales, horses, and sheep are all placental mammals. That means their young develop inside the mother's womb until they are fully formed.

Young are attached to the wall of the womb by a **placenta**. Food and oxygen pass through the placenta from the mother's blood.

Baby rabbits are born below ground, in fur-lined nests.

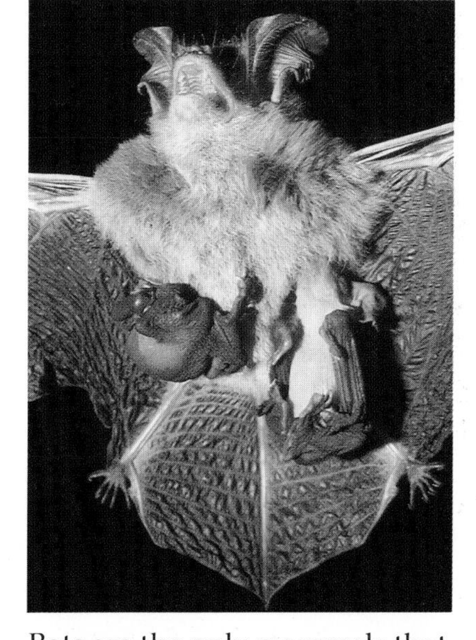

Bats are the only mammals that can fly. A baby bat clings to its mother until it is about two weeks old. Then she hangs it upside down in a safe place while she flies off to find food.

Sea lions are placental mammals too. The pups are suckled until they are about three months old.

Questions

1 Name two things that help birds fly.

2 Name a bird that can't fly.

3 What do human mothers and kangaroo mothers have in common?

4 What are monotremes? Give two examples.

5 Think of two reasons why pouches might be better than eggs, for developing babies.

6 Write down ten examples of placental mammals.

7 Think of three advantages of a mother carrying babies inside her rather than laying eggs.

1·9 Plants

There are two kinds of plants:

1 **Plants without seeds.** These are mosses, liverworts, and ferns. They produce **spores** which grow into new plants.

2 **Seed plants.** These are conifers and flowering plants. They produce **seeds** which grow into new plants.

Plants without seeds (non-flowering plants)

Mosses and liverworts. These usually grow in damp places. Their spores grow in **capsules** on the end of stalks.

Moss plants have stems and tiny leaves. Over 600 different kinds grow in Britain. This one is called bank hair moss.

Moss capsules. When they are ripe their lids drop off, releasing spores which are scattered by the wind.

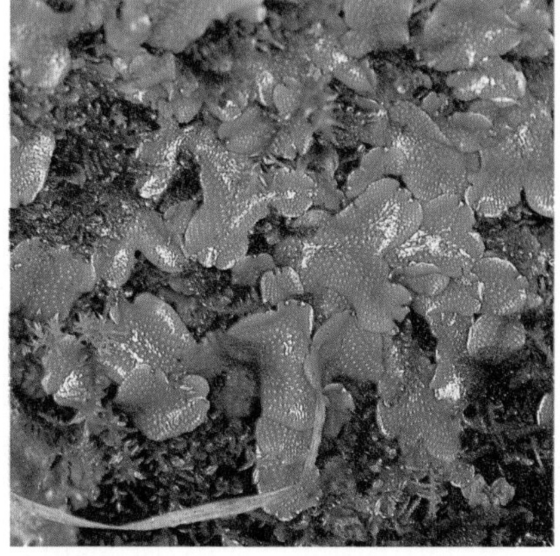

Unlike mosses, many liverworts have no distinct stems. They look like leaves growing flat on the ground.

Ferns. These usually grow in woods and other damp shady places. Their spores grow in capsules attached to the back of the leaves.

Britain has over 50 different kinds of fern. This one is called hard fern. It is common on moorland and heath.

This kind of fern is called bracken. It can grow nearly two metres tall. It is very common on moorland.

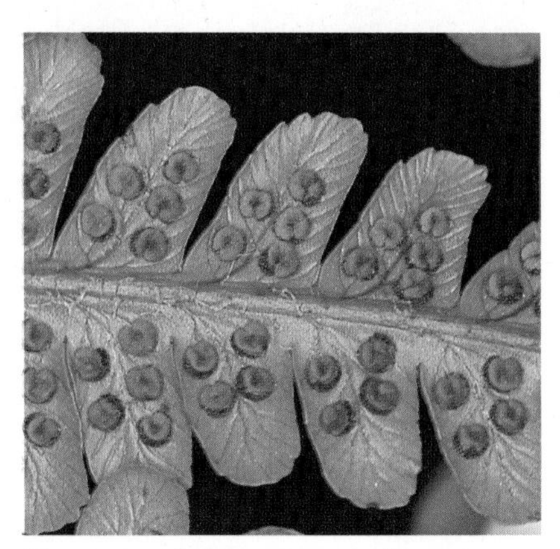

Groups of spore capsules on the back of a fern leaf. Each group is protected by a tiny orange-brown scale leaf.

Seed plants

Conifers. Pines, spruces, larches, cedars, and firs are conifers. They produce male and female cones.

The male pine cone produces pollen, which is carried by the wind to a female cone.

The fertilized female cone grows seeds under its flaps, which open when they are ripe.

A horsechestnut tree in flower. Its hard brown seeds are used in the game of 'conkers'.

Flowering plants. Cabbages, horsechestnut trees, and grasses are all flowering plants. They have flowers with reproductive organs which produce seeds. There are two kinds of flowering plants, **monocotyledons** and **dicotyledons**.

Monocotyledons, like these daffodils, have long thin leaves with parallel veins.

Dicotyledons, like this water lily, have broad leaves with a network of veins.

Questions

1. What are capsules, cones, and flowers for?
2. Which of these plants have capsules? Which have cones? Which have flowers?
   roses, bog moss, Christmas trees, bracken ferns.
3. Which of these are monocotyledons, and which are dicotyledons?
   buttercups, tulips, grasses, apple trees.
   Explain your answer.
4. Name the two main groups of plant.

Questions

1 All living things *move, have senses, feed, respire, excrete, reproduce,* and *grow.*

Write out these seven features as headings across a page. Then sort the list below into seven groups under the headings.

For example, photosynthesis would come under *feed*.

fins	seeds
photosynthesis	eggs
kidneys	taste
babies	sunlight
energy	oxygen
muscles	lungs
eyes	mating
root growth	leaf-fall
leaves	seedlings
breathing	flowers

2 **a)** Write down three headings: *Animals only, Plants only* and *Both animals and plants*.
b) Now sort the list below into three groups under these headings.

excretion
photosynthesis
movement from place to place
growth throughout life
respiration
eating other living things
growth towards light
growth stops when adult size is reached

3 Look at each statement below. Which animal or group of animals does it describe?

a) They have three pairs of legs.
b) They move about on a large slimy foot.
c) They have a thick, hard cuticle, and two pairs of antennae.
d) They have tentacles and a beak like a bird.
e) They carry their babies in a pouch.
f) The body is like a bag, with a mouth and tentacles at one end.

4 Look at each statement below. Which group of plants does it describe?

a) Spore capsules grow on the backs of their leaves.
b) Grasses, sycamore trees and tulips are examples.
c) Their seeds grow inside cones.

5 The drawings below show some living things. Which of them:

a) is a vertebrate?
b) is an invertebrate?
c) is a mammal?
d) is an insect?
e) makes food by photosynthesis?
f) lives by making things decay?
g) reproduces with seeds?
h) reproduces with spores?
i) suckles young on milk?
j) has compound eyes?

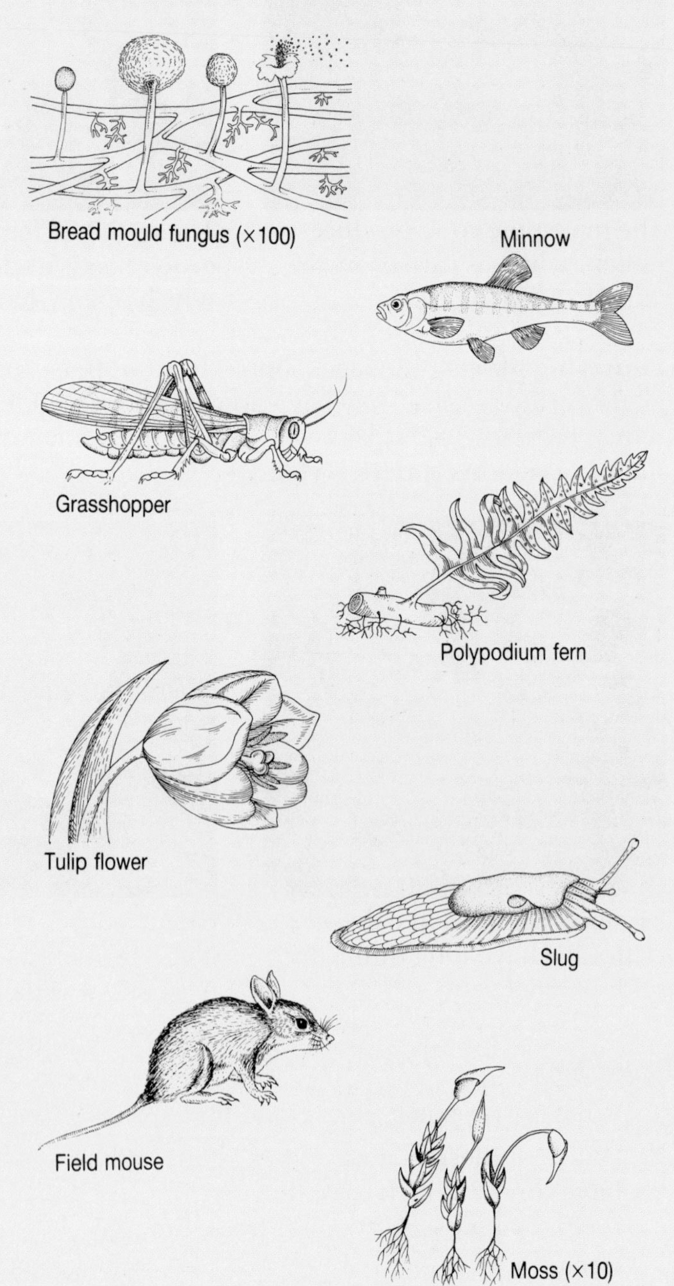

1 Growing algae and protozoa

If your school has an aquarium you may have noticed that the glass gets covered with green scum. The scum is made of algae and protozoa. To see what these look like under a microscope, here is what to do.

a) Tie a microscope slide to a piece of thread, and hang it inside the aquarium.
b) After a week take the slide out, wipe one side dry, and put it under a microscope, with the wet side up.
c) Look at the slide under low, and then high magnification. Your teacher will show you how to do this. You will see many types of algae and protozoa clinging to the glass. You may also see protozoa such as *Paramecium* swimming around (page 8).

2 Looking at worms

You can make a wormery from sheets of perspex, as shown in the diagram below. The sheets should be about 30 cm square.

a) Fill the wormery with different types of soil. Add the soil carefully so that you can see the separate layers. Then sprinkle a little water on it to moisten it.
b) Now move about ten earthworms into your wormery. Scatter some food on the soil for them. For example you can use chopped cabbage and lettuce leaves, grass cuttings, and chopped carrot.
c) Watch to see what the worms do.
d) Now put the wormery in the dark, or cover it with a blanket. Look at it every day. Remove any decayed food and replace it with fresh food.
e) Watch out for any changes to the layers of soil in your wormery. What do you notice?
f) After studying your wormery, write down some ways in which worms keep soil fertile.

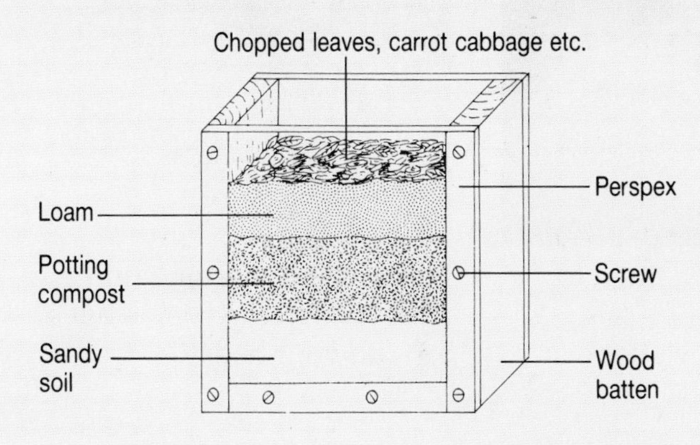

3 Looking at slugs

You can keep slugs or snails in a box for a few days. Use a large plastic box such as an icecream container.

a) Punch holes in the lid of the box, to let air in.
b) Put a layer of moist (not wet) soil, about 5 cm deep, on the bottom of the box. Add some moist dead leaves or a piece of tree bark so that the slugs have somewhere to hide. Put in a jar lid as a food dish.
c) Put two slugs of the same type in your sluggery. Feed them on pieces of lettuce or cabbage leaf, apple, pear, banana, beans, peas, and bread.
d) Watch how each slug moves about on its slimy 'foot'.
e) Watch how the slugs use their tentacles. The long tentacles are eyes. The short ones are for smelling and tasting.
f) Watch how the slugs eat. They grind up food with their tongues which have a rough surface like a file.
g) Each slug has both male and female sex organs. Your slugs may mate and lay batches of pearly white eggs. Put some eggs on a damp paper towel and look at them through a hand lens. You will see tiny developing slugs inside.

4 Studying caterpillars

a) Search for some caterpillars on plant leaves.
b) Take a sprig of the plant, complete with caterpillars and put it in an insect cage, as in the diagram.
c) Now find out the names of the caterpillars and the plant. (You can look them up in books.)
d) Use a hand lens to watch how the caterpillars eat. You will see their jaws move.
e) Remove the plant sprig when the leaves have been eaten. Use a large paint brush to move the caterpillars to a fresh sprig.
f) If you watch the caterpillars regularly, you will see them turn into pupae. What do the pupae look like?
g) Move the pupae into separate cages until the adults emerge. What do these adults look like? Take them to the place you found the caterpillars, and set them free.

2·1 What are cells?

A house is built of bricks. In the same way, animals and plants are built of **cells**.

Your skin, your bones, your muscles, and your brain are all made of cells. There are around a hundred million million cells in your body. They are very tiny. At least ten cells would fit side by side across this next full-stop.

A cell from the lining of a human cheek, coloured to show up the different parts. What do you think the orange blob is?

Inside an animal cell

Your cells are not very different from the cells of a frog or a cat or a giraffe. In fact all animal cells have these parts:

A cell membrane. This is a thin skin around the cell. It lets some things pass through, but stops others.

Cytoplasm. This is a jelly containing hundreds of chemicals. Lots of chemical reactions go on in it. It fills the cell.

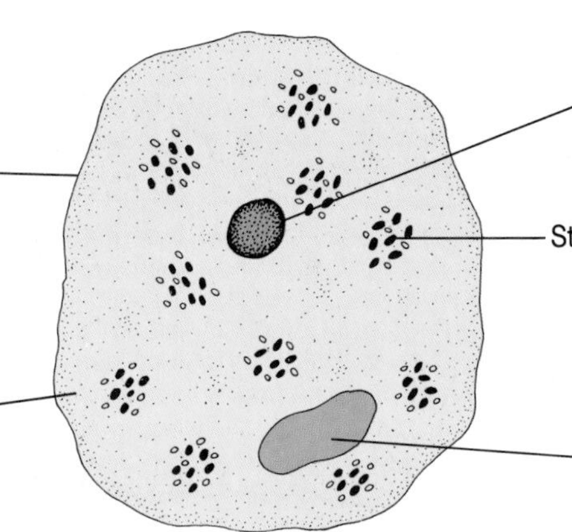

A nucleus. It controls what a cell does, and how it develops.

A vacuole. This is a space within the cell containing air, liquids, or food particles. Animal cells usually have several small vacuoles.

Cells are not all the same shape. There are about twenty different types of cell in your body, all doing different jobs. Below are photographs of three types of cell.

Red blood cells are disc shaped. Their job is to carry oxygen round the body.

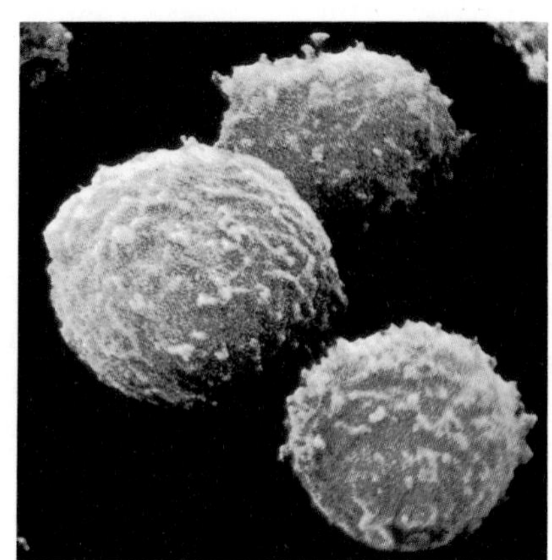

White blood cells, stained to show up clearly. They can change shape. They attack germs.

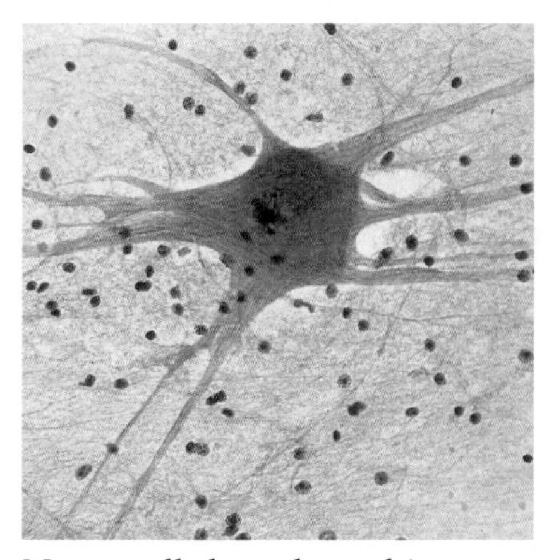

Nerve cells have long thin fibres which carry 'messages' around your body.

Inside a plant cell

All plant cells have these parts:

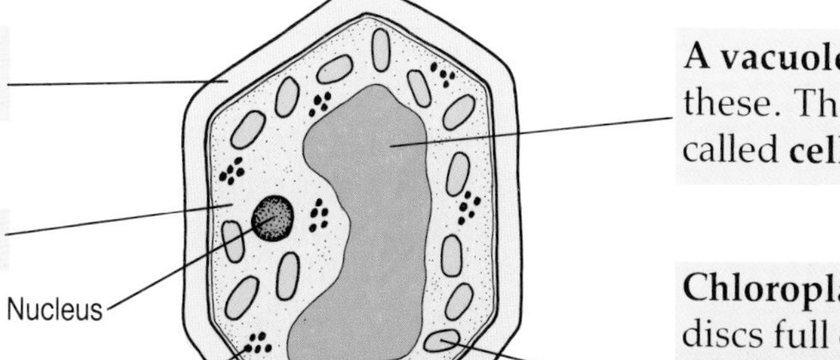

Unlike animal cells, there are only a few different shapes of plant cells. This is because there are not so many different jobs for them to do.

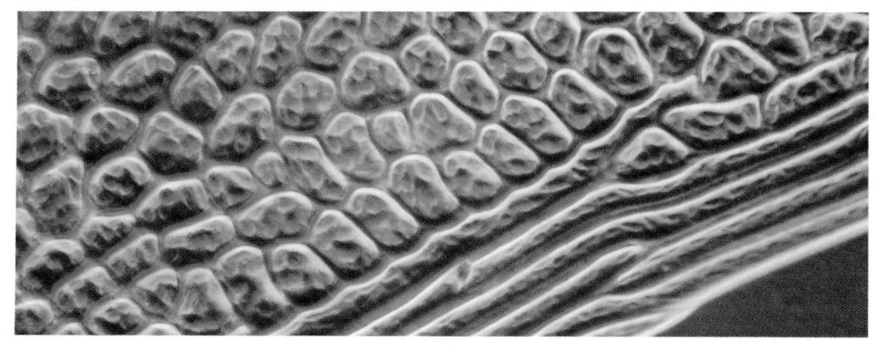

Leaf cells. They look firmer than animal cells because of their cell walls. The tiny green blobs are chloroplasts. There are several in each cell.

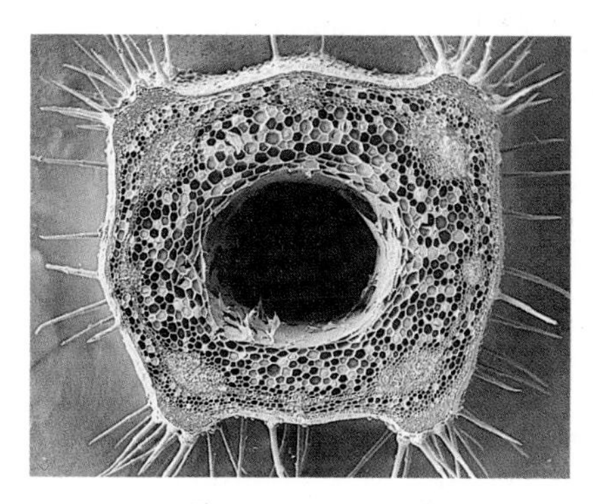

A slice through a nettle stem. Some cells form tubes to carry water, others to carry foods.

How plant cells are different from animal cells

Plant cells	Animal cells
1 have cellulose cell walls	**1** do not have cellulose cell walls
2 have chloroplasts	**2** do not have chloroplasts
3 always have a vacuole	**3** sometimes have a vacuole
4 have a few different shapes	**4** have many different shapes

Questions

1. Name three different kinds of cell in your body.
2. What is cytoplasm?
3. What does a cell nucleus do?
4. What are chloroplasts. What do they contain? What do they do?
5. How are plant cells different from animal cells? List four differences.
6. List four things found in *both* plant cells and animal cells.

2·2 How living things grow

You started life as a single cell – a fertilized egg cell. You grew because that cell divided to make two cells, these divided to make four, and so on. This is called **cell division**. It is how all living things grow.

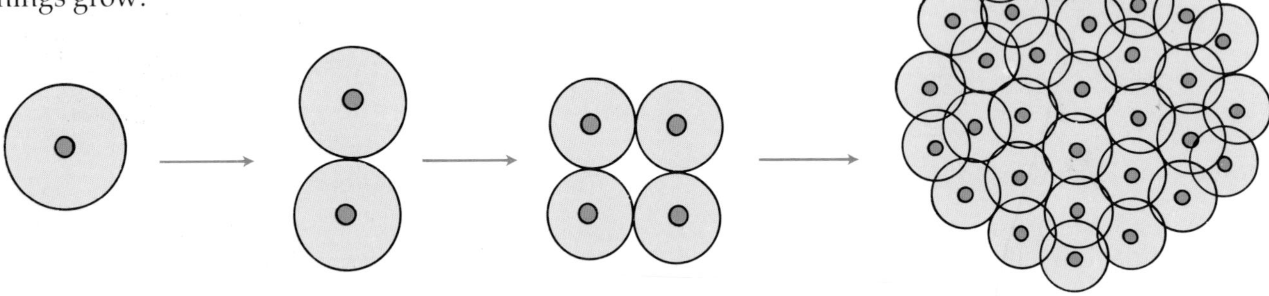

A fertilized egg cell divides to make *two* **daughter** cells, which are identical.

These divide to make *four* identical cells, which divide again and again to make a ball of cells.

Tissues

Some cells in the ball grow and change shape to do a particular job – they become **specialized**. Cells that do the same job group together to form **tissues**.

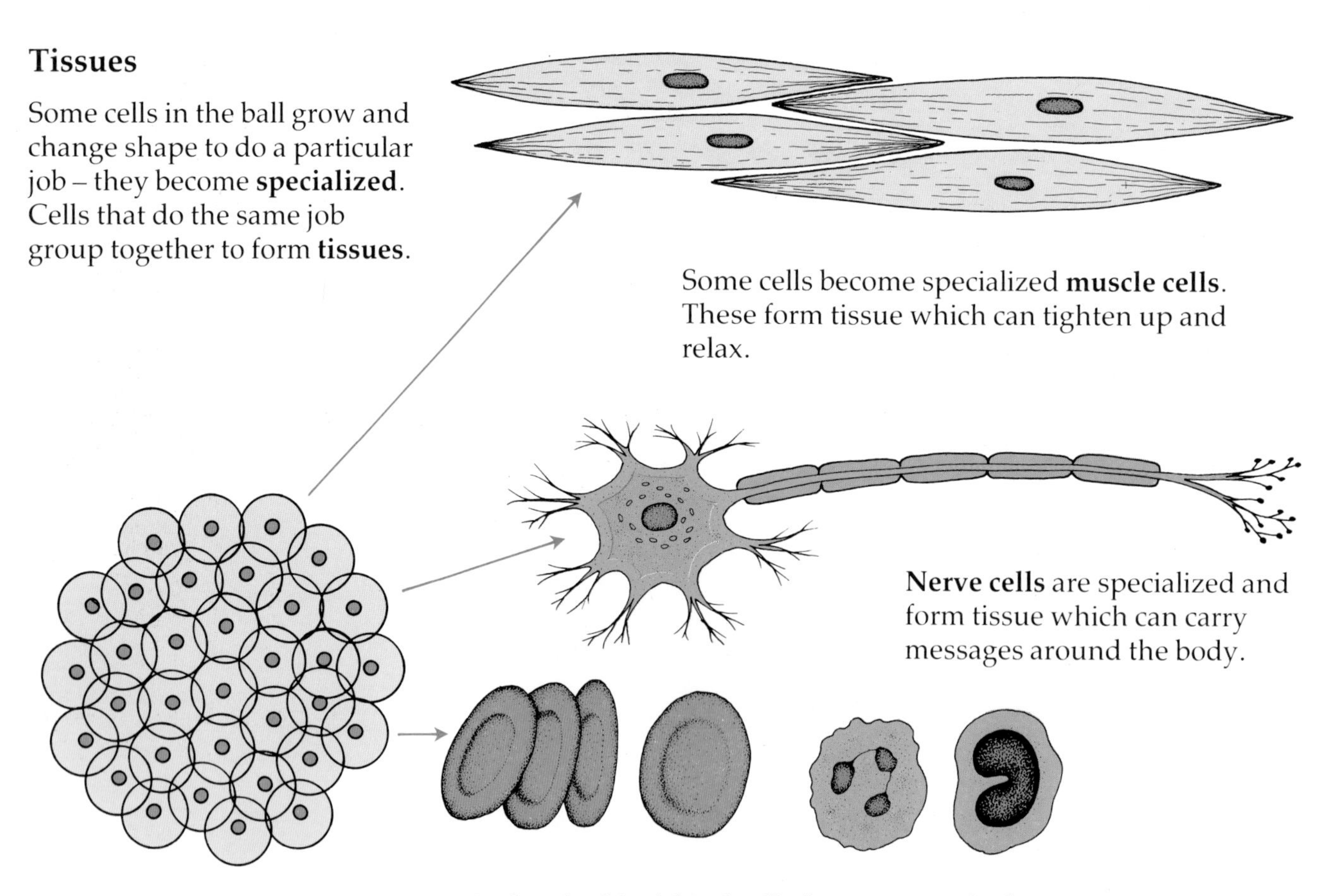

Some cells become specialized **muscle cells**. These form tissue which can tighten up and relax.

Nerve cells are specialized and form tissue which can carry messages around the body.

Red and white blood cells form tissue which can carry oxygen and kill germs. This tissue is called **blood**.

Organs

Different tissues combine to make **organs**.

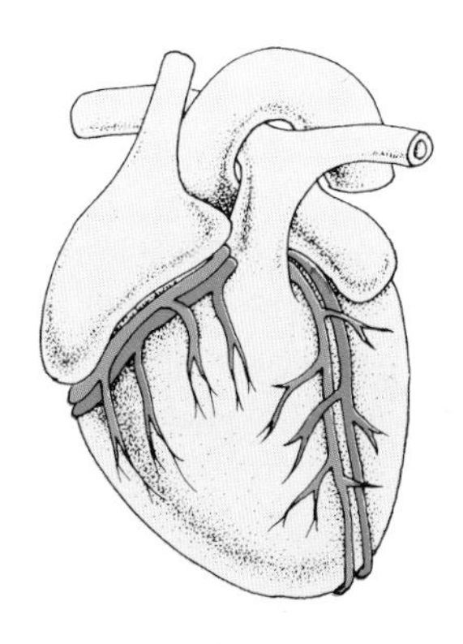

The **heart** is an organ which pumps blood around the body.

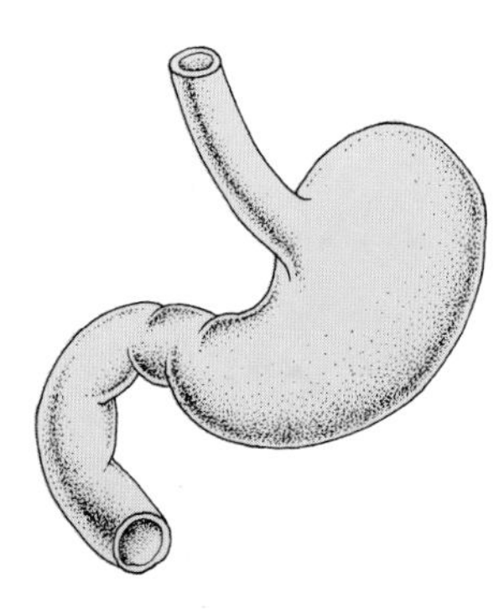

The **stomach** is an organ which digests food.

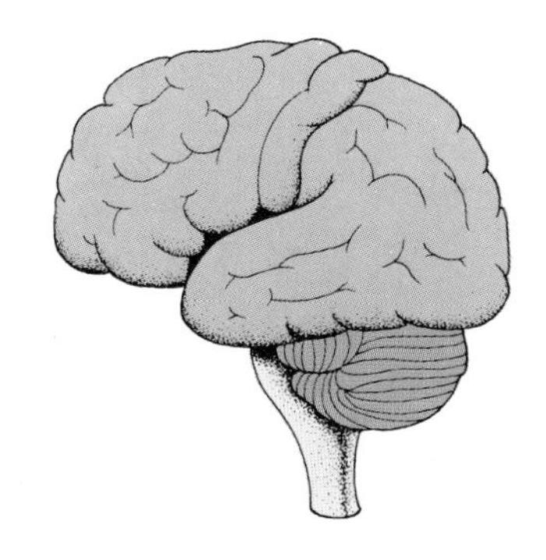

The **brain** is an organ which controls parts of the body.

Organisms and organ systems

You are an **organism**. So is a cat and so is a bird. An organism is made up of many different organs. Some of its organs work together to form **organ systems**.

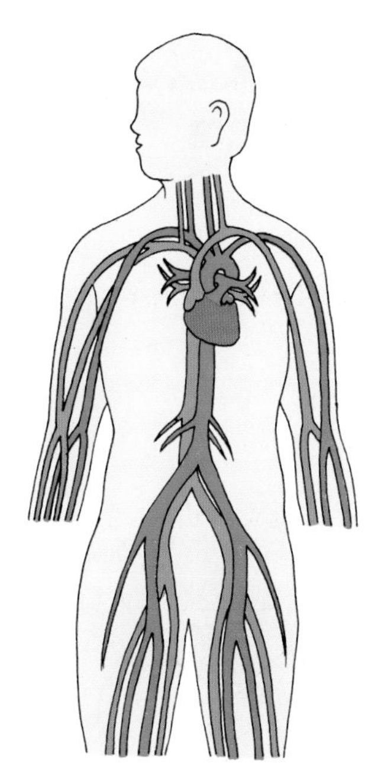

The **circulatory system** is made up of the heart and blood vessels.

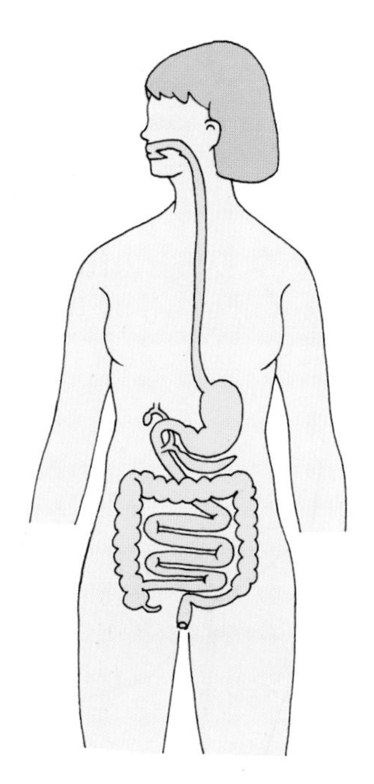

The **digestive system** is made up of the gullet, stomach, and intestine.

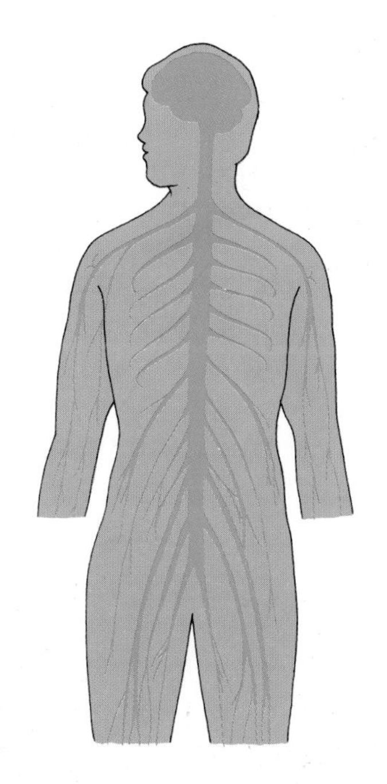

The **nervous system** is made up of the brain, spinal cord, and nerves.

Questions

1 What is a tissue? Give two examples.

2 What is an organ? Give five examples.

3 What is an organ system? Give two examples.

4 Even when you are fully grown some cells carry on dividing. Explain why.

5 Explain what a specialized cell is.

2·3 In and out of cells

Substances pass in and out of cells as tiny particles called **molecules**.

Molecules are the smallest particles of a substance. The molecules in liquids and gases are never still. They keep moving and bumping into each other.

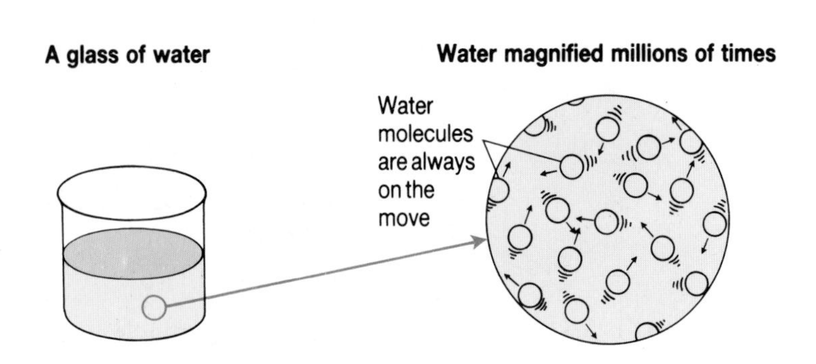

An experiment to show that molecules move

A drop of ink is added to water. The ink spreads through the water even though it is not stirred.

The ink spreads because ink molecules move into the spaces between water molecules, and water molecules move into the spaces between ink molecules.

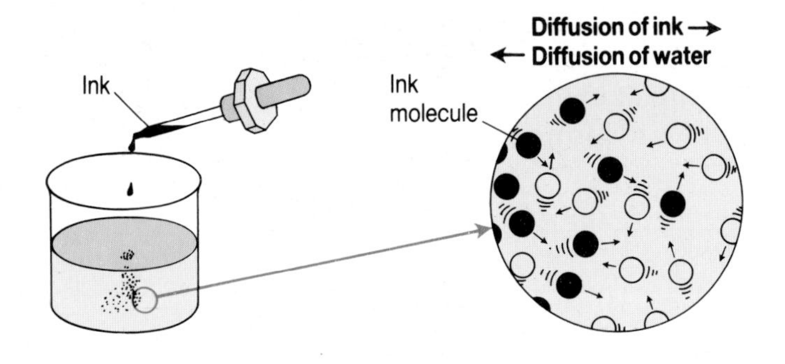

The movement of molecules so that they mix is called **diffusion**. Molecules diffuse from where they are plentiful to where they are less plentiful.

Diffusion in and out of cells

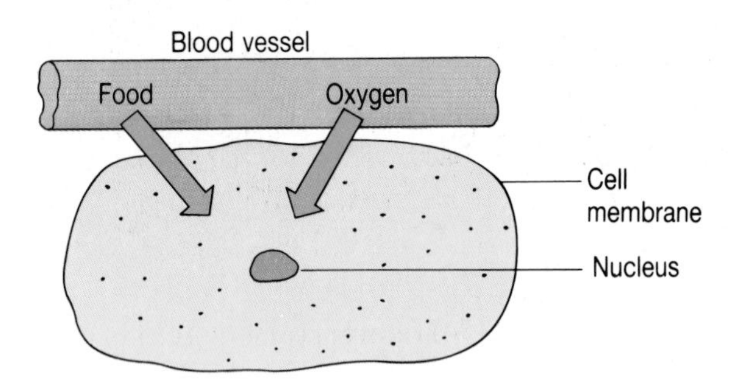

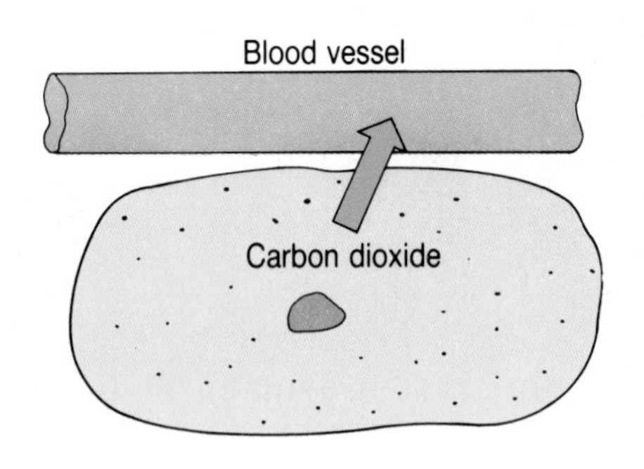

Human body cells need a constant supply of food and oxygen to stay alive and do their jobs. Food and oxygen are carried in the blood, so they diffuse from the blood into each cell.

As the cell uses food and energy it produces carbon dioxide as waste. This waste must be removed before it poisons the cell. Carbon dioxide diffuses from the cell into the blood which carries it away to the lungs to be breathed out of the body.

Osmosis

Osmosis is a special kind of diffusion. It happens when a membrane has tiny holes in it which let water molecules pass through but stop larger molecules, like sugar. A membrane like this is called **semi-permeable**. This experiment shows osmosis:

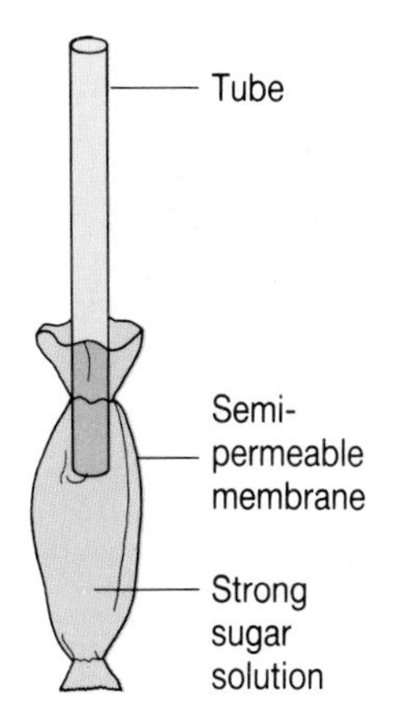

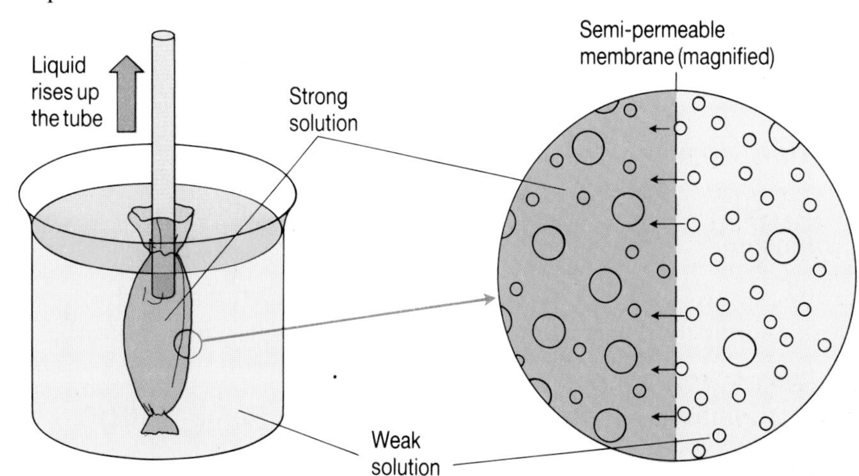

A semi-permeable membrane is tied to a tube. It is then filled with strong sugar solution.

The membrane is stood in a weak sugar solution. Soon liquid starts to rise up the tube.

The liquid rises because water molecules diffuse through the membrane from the weak solution to the strong one. But sugar molecules cannot diffuse like this because they are too big to pass through the membrane.

When a weak solution is separated from a strong one by a semi-permeable membrane, water always flows from the weak solution to the strong one. This diffusion of water is called **osmosis**.

Osmosis in plant cells

Water moves from cell to cell in plants by osmosis. The cell membrane of a plant cell is semi-permeable. So if a cell containing a weak solution is next to a cell with a stronger solution, water moves by osmosis from the weak to the strong solution, as shown in this diagram.

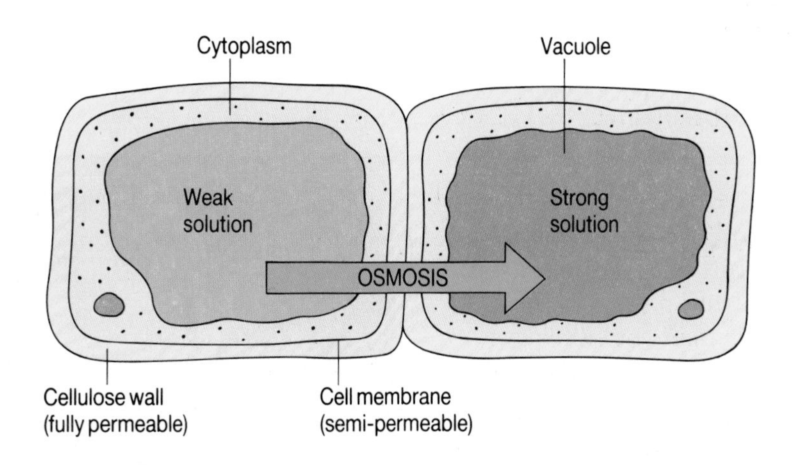

Questions

1. Why does ink move through water even without being stirred?
2. How does oxygen move from blood into body cells?
3. Osmosis is the movement of ______ from a ______ solution to a ______ solution through a ______ membrane.
4. What is a semi-permeable membrane?

2·4 Cell division

You started life as a fertilized egg cell, smaller than a full stop. You grew because this cell divided millions and millions of times to make new cells.

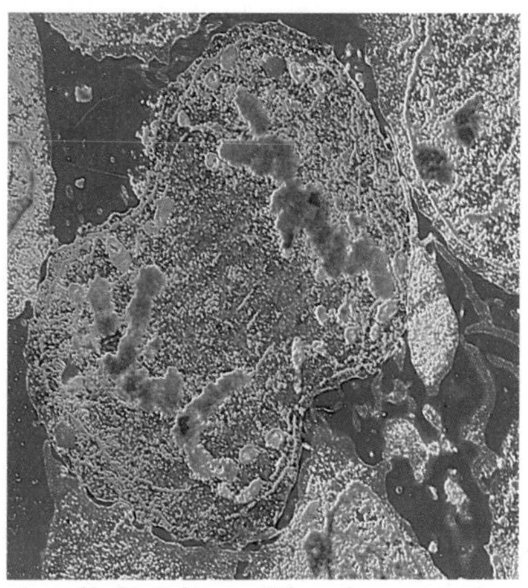
A cell dividing. The deep pink parts are the chromosomes.

Chromosomes

When a cell is ready to divide, long thin threads called **chromosomes** appear in its nucleus.

Chromosomes contain chemicals which control what a cell does. These chemicals also contain all the instructions needed to build a whole new organism from a single fertilized egg cell.

When a cell divides it makes a new set of chromosomes so that these instructions can be passed to the daughter cells.

The chromosomes are always in pairs.

How a cell divides

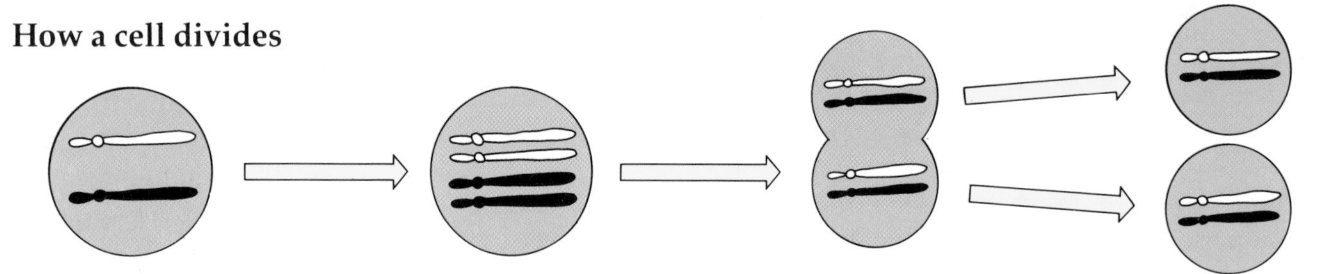

This cell has only two chromosomes. (Human cells have **46** chromosomes altogether.) Each chromosome splits into two, making a second set of chromosomes.

The *cell* divides in two. Each cell gets a *full set* of chromosomes. The two new cells have the same number of chromosomes as the first.

This kind of cell division is called **mitosis**. All the cells in animals and plants, *except* sex cells, are made by mitosis.

Mitosis produces:
1 the cells needed to make an adult organism from a fertilized egg
2 the cells needed to heal cuts, wounds, and broken bones
3 the cells that replace dead skin cells, and worn out red blood cells.

Mitosis produces the cells that let a seed grow into a plant.

Mitosis produces new bone cells which mend a broken arm.

How sex cells are produced

Some of the cells in your body are sex cells, or **gametes**. If you are male they are called **sperms** and are made in your **testes**. If you are female they are called **ova** or eggs, and are made in your **ovaries**. Sex cells are made by a different kind of cell division.

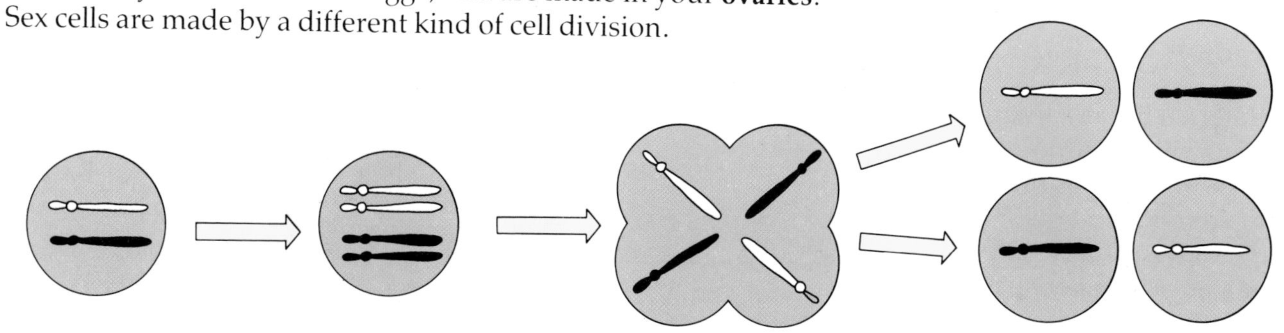

This cell is drawn to show only two chromosomes. Each chromosome splits into two, making a second set of chromosomes.

The cell divides to form *four* sex cells. Each sex cell gets only *half as many* chromosomes as the first cell. (Human sex cells have only **23** chromosomes.)

This kind of cell division is called **meiosis**. Meiosis produces sex cells in humans and other animals, and in plants.

Fertilization

At fertilization a male sex cell joins up with a female sex cell to make a fertilized egg cell called a **zygote**.

This drawing shows fertilization in humans.

In human fertilization a sperm with **23** chromosomes joins an ovum with **23** chromosomes to make a zygote with **46** chromosomes.

Fertilization happens in the same way in other animals, and in flowering plants. In flowering plants, the fertilized sex cell grows into a seed.

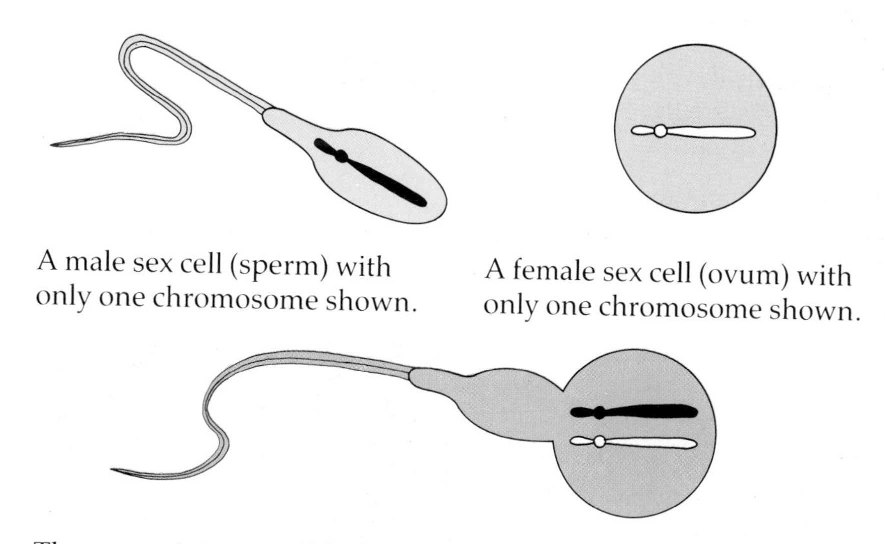

A male sex cell (sperm) with only one chromosome shown.

A female sex cell (ovum) with only one chromosome shown.

The sperm joins up with the ovum. This produces a zygote with the *full number of chromosomes*. This grows by mitosis into an adult.

Questions

1 Which kind of cell division:

a) halves the chromosome number?

b) produces cells to repair wounds?

c) produces sperms?

d) produces an adult from a zygote?

2 How are zygotes produced?

2·5 Heredity and variation

Heredity

You started life as a fertilized egg cell with 46 chromosomes. 23 of these chromosomes came from your father, and 23 from your mother. This is why you have some characteristics of your father and some of your mother.

All children inherit certain characteristics from their parents. The study of **inherited characteristics** is called **heredity**.

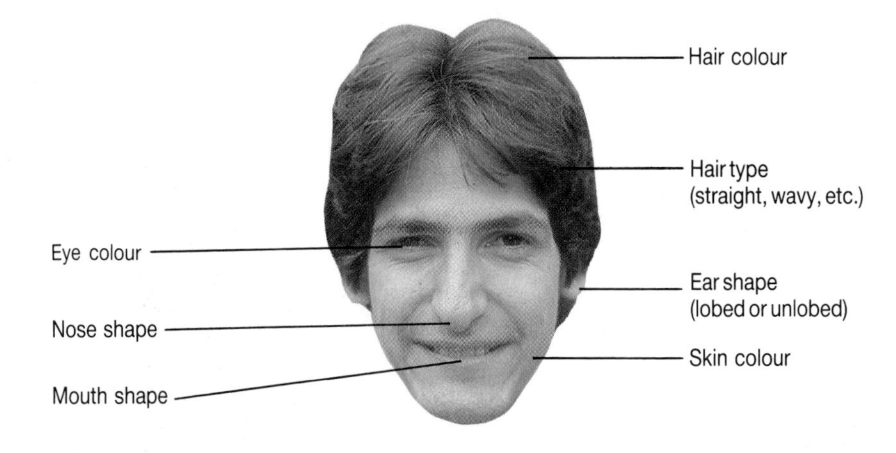

Inherited characteristics

The photograph above shows some of the characteristics children inherit from their parents. The development of these characteristics is controlled by chromosomes.

The children in the family below inherited some characteristics from each of their parents. Try to sort out which characteristic came from which parent.

Father has red curly hair, brown eyes, a snub nose, and a cleft chin.

Mother has straight black hair, blue eyes, a long thin nose, and a pointed chin.

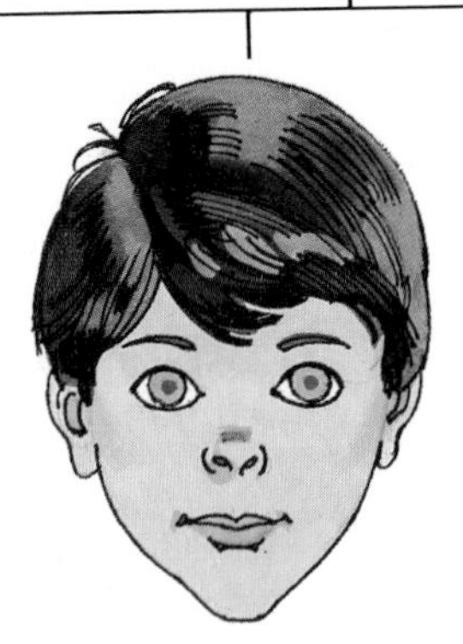

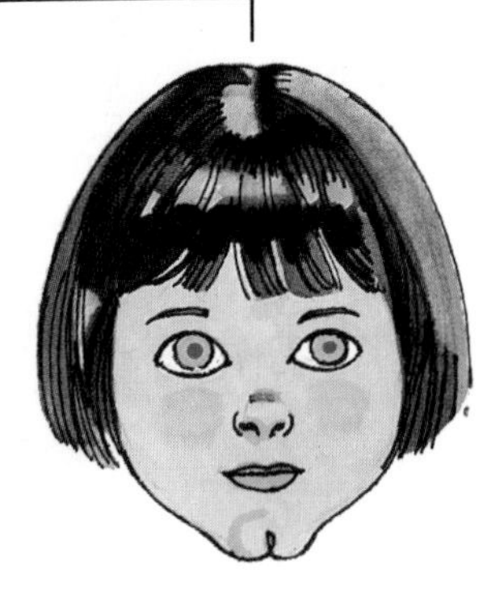

Acquired characteristics

You may know how to swim, or rollerskate, or speak French. Or you may have a scar from a cut. These are **acquired characteristics**. You pick them up (acquire them) as you go through life. You are not born with them, and you cannot pass them on to your children.

Variation

No two people are exactly the same. Even identical twins are different in some ways.

People are different heights and weights. Their hair and eyes are different colours and their faces have different shapes. Their eye-colour and hair-colour show **variation**.

Continuous variations

These people are arranged in a line from the shortest to the tallest. Their height shows **continuous variation**. It varies from short to tall with many small differences in between.

Intelligence also shows continuous variation. Can you think of other variations which are continuous?

People's height shows continuous variation.

Discontinuous variation

People can either roll their tongues or they cannot. This is an example of **discontinuous variation**. You can either do it or you can't. There is no 'in between' state.

Blood groups show discontinuous variation. You belong to only one group: A, B, AB, or O. Find out what blood group each member of your class belongs to.

Can you think of any other variations that are discontinuous?

Tongue rolling. Either you can. . .or you can't.

Questions

1 Write down eight inherited characteristics.

2 Write down six acquired characteristics.

3 Intelligence shows continuous variation. What does this mean?

4 Colour blindness shows discontinuous variation. What does this mean?

5 Write down four of your own acquired characteristics.

2·6 Chromosomes and genes

You inherit characteristics from your parents through their chromosomes.

A chromosome has small parts called **genes** all the way along it. The genes are made of a chemical called **DNA**. Genes control the development of inherited characteristics. For example, there are genes which control eye colour, hair colour and skin colour.

What happens to genes during fertilization

A sperm has 23 chromosomes with genes from a man. An ovum (egg) has 23 chromosomes with genes from a woman. During fertilization the sperm and ovum join together.

Each chromosome from the sperm then pairs up with a matching chromosome from the ovum. This brings the two sets of genes together. The genes for hair colour pair up, the genes for skin colour pair up, and so on. *Genes always work in pairs.*

The genes in a pair may be identical. For example, they may both produce black hair. But if one is a gene for black hair and the other a gene for blond hair they are in competition.

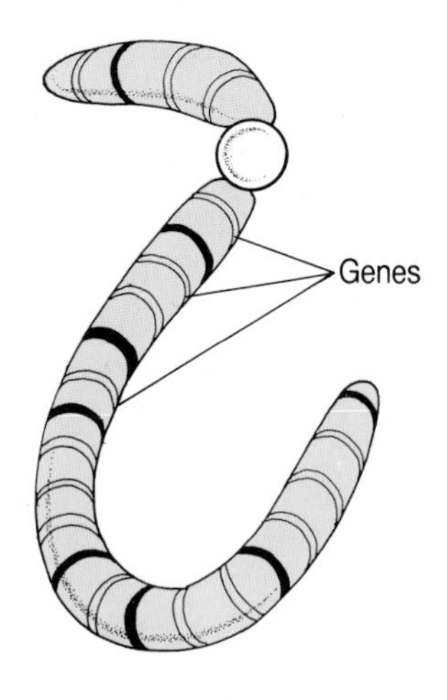

This is what a chromosome would look like if you could see the genes.

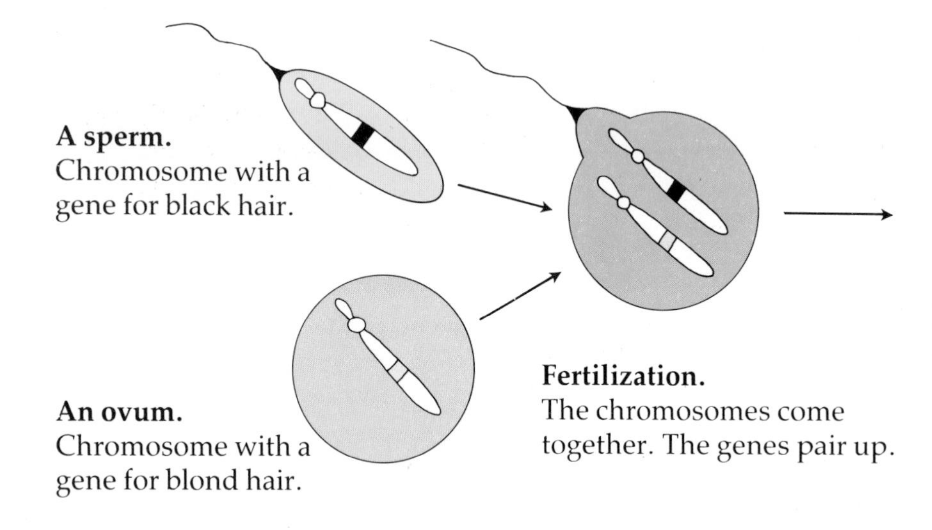

The child has black hair. Only the gene for black hair has worked.

Dominant and recessive genes

The child has black hair because the gene for black hair is more powerful than the gene for blond hair. It **dominates** the gene for blond hair and produces the final hair colour. Genes which dominate other genes are called **dominant genes.**
Genes which are dominated are called **recessive genes.** When a dominant gene pairs up with a recessive gene, the dominant one produces the final effect.

Genes in action

In diagrams, genes are usually shown by letters. Capital letters are used for dominant genes and small letters for recessive genes. In the diagram below, H is the gene for black hair and h is the gene for blond hair.

Mother's cells contain two genes for black hair (HH) so she has black hair.

Father's cells contain two genes for blond hair (hh) so he has blond hair.

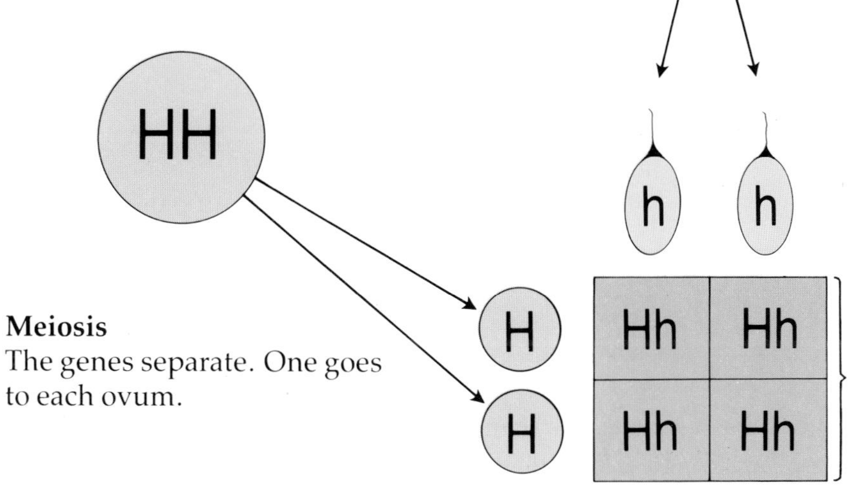

Meiosis
The genes separate. One goes to each sperm.

Meiosis
The genes separate. One goes to each ovum.

Zygotes
All have the genes Hh. So all the children will have black hair, because the gene for black hair is dominant.

Homozygous and heterozygous

A person with two identical genes for a characteristic is pure bred or **homozygous** for that characteristic. (HH is a pure bred gene.) A person with two different genes (one dominant and one recessive) for a characteristic is **hybrid** or **heterozygous** for that characteristic. (**Homo** means the same and **hetero** means different.)

Questions

1 Which parts of chromosomes control inherited characteristics?

2 What are genes made of?

3 In the gene pair Hh (for black and blond hair) which gene is dominant?

4 If E is the gene for brown eyes, and e is the gene for blue eyes, what eye colour will the following gene pairs give?
a) EE **b)** Ee **c)** ee

5 Look at question 4 and pick out a gene pair which is heterozygous, and a gene pair which is homozygous.

2·7 More about chromosomes

On the previous page all the children had dark hair, even though their father was blond. This happened because they inherited a dominant gene (H) and a recessive gene (h). They were **hybrids**, or **heterozygous** (Hh) for hair colour.

This does not mean that blond hair has been lost altogether. Look what happens when two hybrids have children.

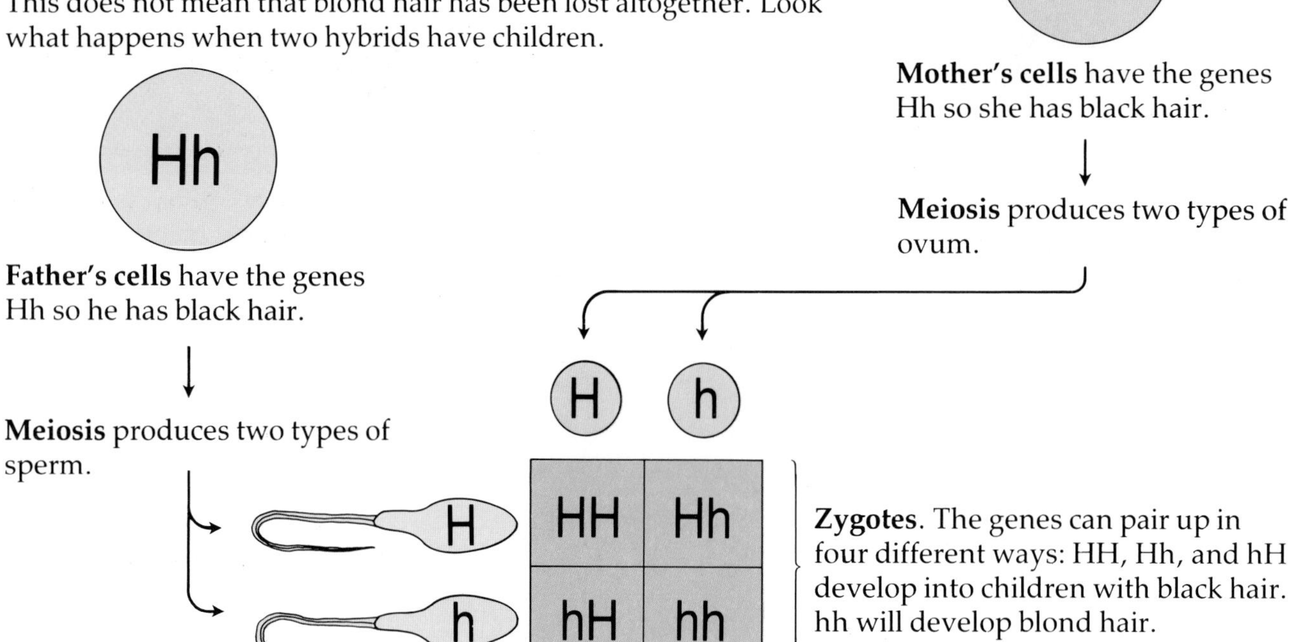

Blond hair appears this time because two recessive genes (hh) can come together. But the children are three times more likely to have black hair than blond hair.

Other dominant characteristics

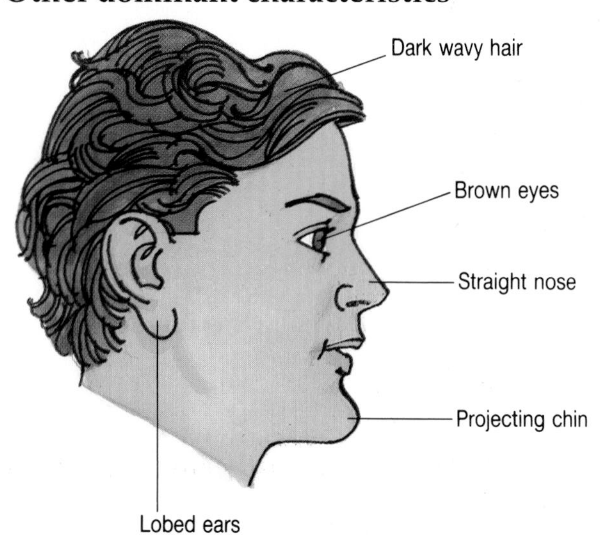

Other recessive characteristics

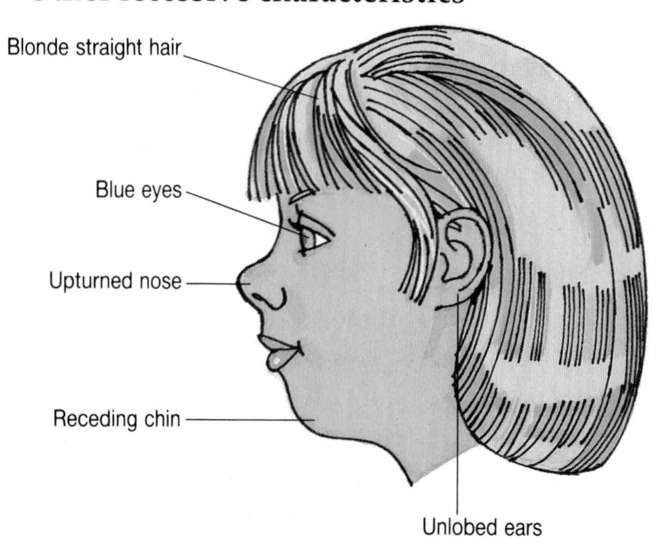

Boy or girl?

Human cells have 46 chromosomes altogether. There are 22 matching pairs. But the last pair does not always match. These two chromosomes are the **sex chromosomes**. They control whether a baby develops into a boy or a girl.

If you are male your cells have an X chromosome, and a smaller Y chromosome. If you are female your cells have two X chromosomes. This diagram shows how sex is inherited.

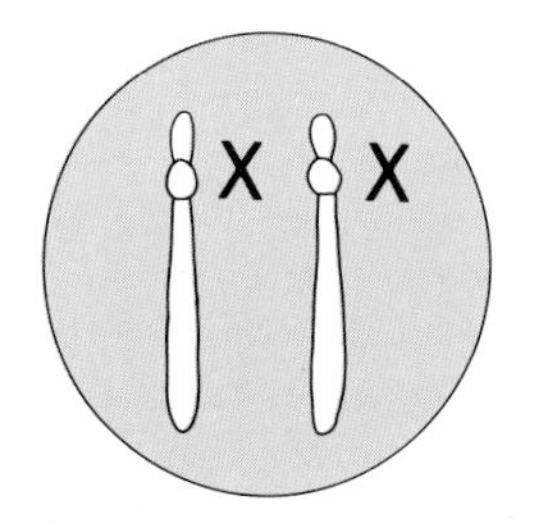

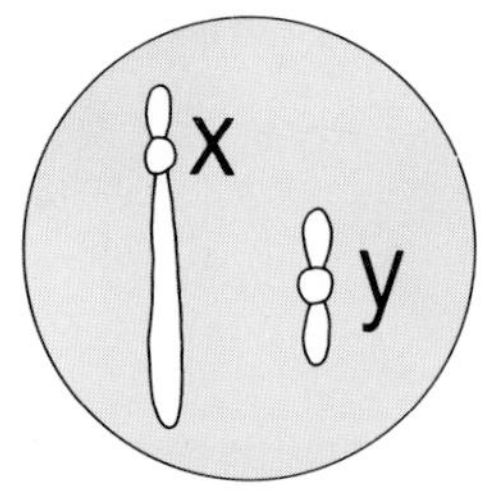

Mother's cells contain two X chromosomes.

Meiosis produces ova with one X chromosome each.

Father's cells contain an X and a Y chromosome.

Meiosis produces equal numbers of X and Y sperms.

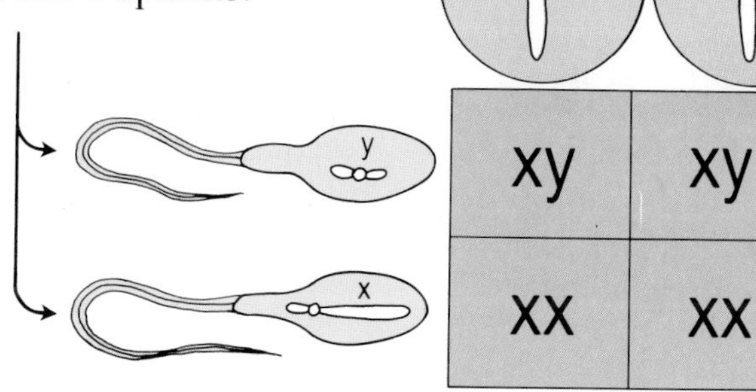

XY zygotes develop into boys.

XX zygotes develop into girls.

Since there are equal numbers of X and Y sperms a child has an equal chance of being a boy or a girl.

This picture shows the 46 chromosomes of a man. They were photographed under a microscope, then cut out and sorted into 22 matching pairs, and a pair of X and Y chromosomes.

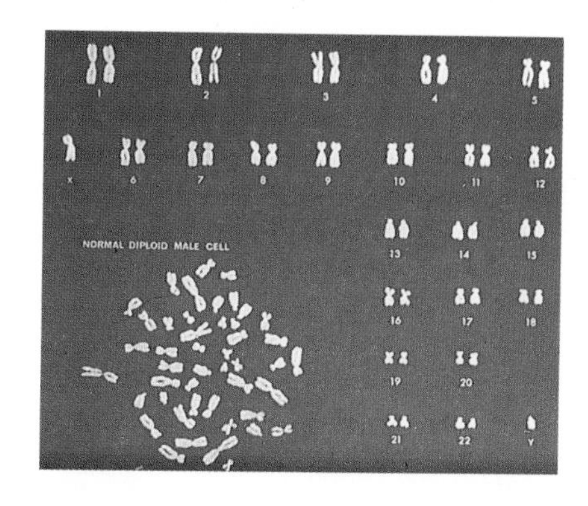

Questions

1 W is the gene for wavy hair, and w is the gene for straight hair. Father has the genes WW. Mother has the genes Ww. Draw a diagram like the one on the previous page to show what type of hair their children could inherit.

2 What do XY zygotes develop into? What do XX zygotes develop into?

3 Why are there roughly equal numbers of boys and girls?

2·8 More about genes

Alleles

Some people have brown eyes. Some have blue eyes or green eyes. Each eye colour is controlled by a different gene.

All the different genes which control a characteristic are called the **alleles** of that characteristic. So the genes for brown, blue, and green eyes are the alleles for eye colour.

Phenotype and genotype

A **phenotype** is a characteristic you can see, such as black hair, an up-turned nose, or the colour of a flower.

A **genotype** is the set of genes which produce a phenotype.

Genotype also means the complete set of genes in an organism. So your genotype is the complete set of genes which made a zygote turn into you.

Flower colour is a phenotype produced by the genotype CC or Cc. White flowers (flowers with *no* colour) have the genotype cc.

Black hair is a phenotype produced by the genotype HH or Hh. Blond hair is a phenotype produced by the genotype hh.

Genes and the environment

Organisms with the same genotype should look the same. But they may look different if they grow in different surroundings or **environments.** The environment can affect the way genes work.

These maize plants have the same genotype. But they look different because they were grown in different soils.

The ones captioned 'Complete' were grown in soil that had everything they needed. The ones captioned 'No nitrogen' were grown in soil that had everything *except* nitrogen. So the different soils have given different phenotypes from the same genotype.

Complete nutrients.

No nitrogen.

Mutations

A **mutation** is a sudden change in a gene or chromosome. A mutation changes the way an organism develops.

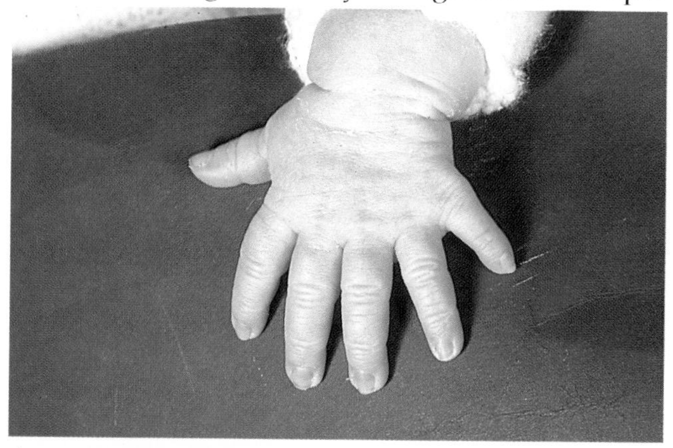

A gene mutation can cause extra fingers and toes to grow. This does not happen very often.

A chromosome mutation can cause a woman to produce an ovum with 24 chromosomes instead of 23. If it gets fertilized the baby has 47 chromosomes instead of 46, and develops **Down's syndrome**. About 1 in 700 babies are affected.

What causes mutations?

Mutations can occur naturally. They can also be caused by x-rays, other radiation, and by some chemicals.

The fruitfly on the left was exposed to gamma radiation, which caused a gene mutation. As a result, its offspring (on the right) has crumpled wings.

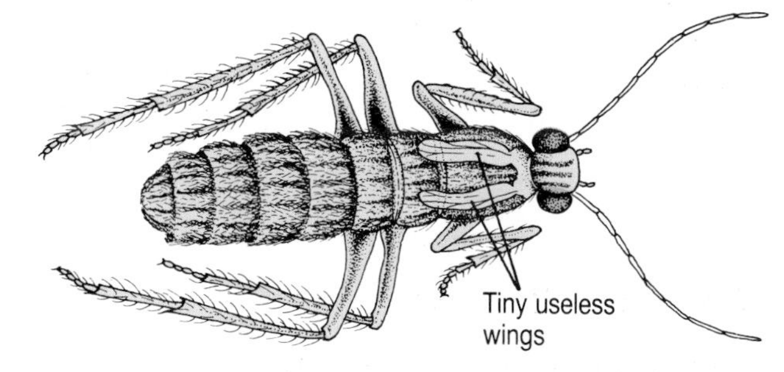

A flightless moth from the Island of Kerguelen in the Indian Ocean. Tiny, useless wings are caused by a gene mutation. Usually the insects don't live as long as their winged brothers and sisters. But Kerguelan is very windy, and the winged insects get blown out to sea. Crawling, flightless insects are safe.

Questions

1 In fruit flies, R gives red eyes and r gives white eyes. What is the phenotype of these genotypes:
a) RR? **b)** Rr? **c)** rr?

2 In mice, B gives dark hair and b gives light hair. What are the genotypes for dark-haired mice?

3 What is a mutation? What causes mutations?

4 Describe a harmful mutation, and a helpful mutation.

2·9 Evolution

Where have the millions of different living things come from? The most likely answer is that they were produced by **evolution.**

What is evolution?

Evolution means change and improvement from simple beginnings. Modern aircraft were evolved by changing and improving the first simple aircraft designs.

Evolution of living things

The first living things appeared about 3500 million years ago. They were no more than bubbles full of chemicals but they could reproduce.

Some of the young were different from their parents. Some of the differences meant they could survive better than their parents. Over billions of years, these changes and improvements led to all the different creatures alive today.

The first powered aircraft evolved to become Concorde.

This diagram shows how vertebrates could have evolved from simple beginnings.

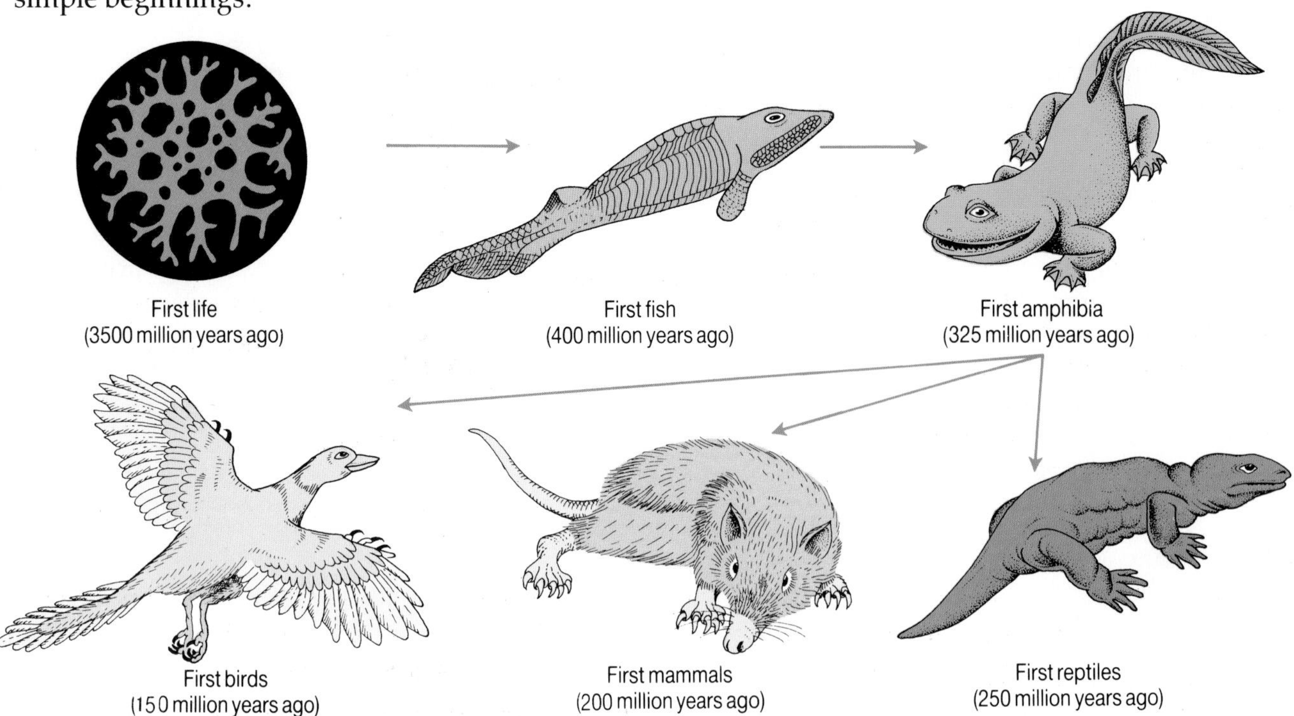

Natural selection

Can you spot the pale Peppered moth? Here its colouring helps to protect it from enemies.

Living things are at risk from disease, predators, and other dangers. If the young are born with differences that make them tougher, or stronger, or better suited to their surroundings, they will be able to live longer and have more babies than their weaker brothers and sisters.

This idea is called **survival of the fittest,** or **natural selection**. Nature *selects* the fittest and strongest for survival by killing off the weaklings. The idea was first put forward by Charles Darwin about a hundred years ago.

Darwin suggested that the young inherit improvements from their parents. In this way a species could keep on improving until it produced a new species.

These photographs show an example of natural selection.

They show dark and light varieties of the Peppered moth. In soot-blackened city areas the light ones are easily seen and eaten by birds, so the dark ones are common. But in clean country areas the dark ones are easily found by birds, so that light ones are more common.

The dark Peppered moth is safer resting on dark surfaces, like this sooty tree trunk. . .

. . . but a pale Peppered moth on a dark surface is easily spotted by birds.

Artificial selection

Animal and plant breeders have proved that selection can change a species. They use the **artificial selection** process.

They select and breed sheep with the longest coats, pigs with the most meat, trees with the juiciest fruit, and plants with the most colourful flowers. So we now have thousands of plants and animals that are very different from their wild ancestors.

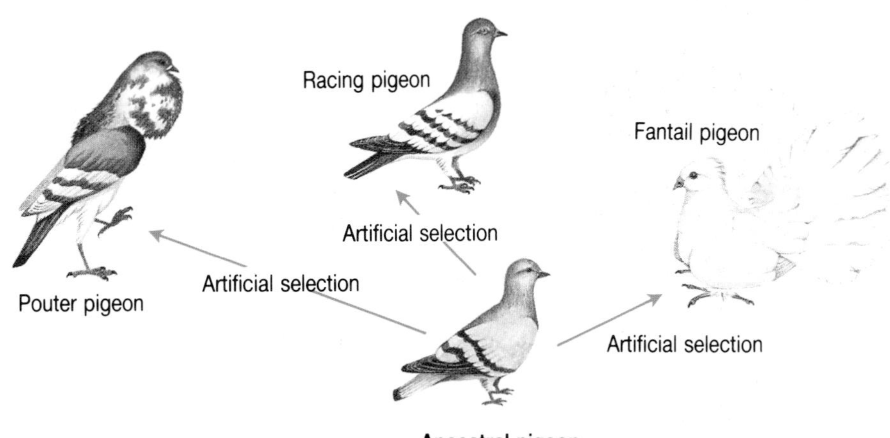

All these different pigeons were produced by artifical selection.

Questions

1. What does evolution mean?
2. What does natural selection mean?
3. How does natural selection make dark Peppered moths more common near factories than light ones.
4. Explain how one species may produce a new species over millions of years.
5. What does artificial selection mean?
6. Give one example of artificial selection.

Questions

1 **a)** Which of the cells drawn opposite is from a plant? Which is from an animal? How can you tell?
b) Parts of the cells have been labelled 1 to 8. Name these parts.
c) Copy down the following descriptions of parts of cells. Then, beside each, write the name and number of the part it describes.
A semi-permeable membrane.
Contains chlorophyll.
Control-centre of a cell.
A bubble of liquid full of chemicals.
Made of cellulose.

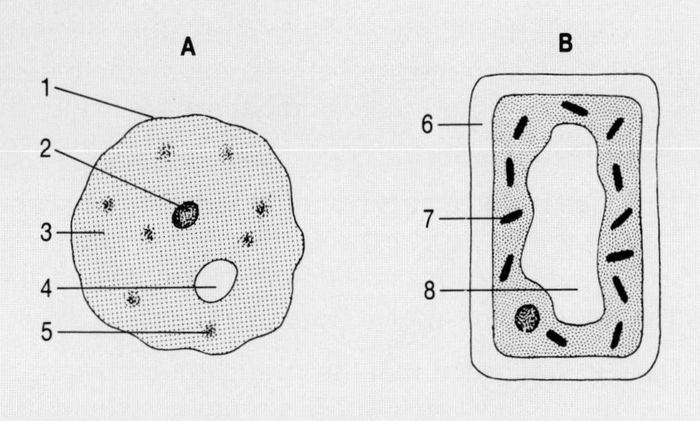

2 **a)** A strong sugar solution was poured into a bag made of semi-permeable membrane.
b) The bag was placed in a beaker of water (diagram A opposite).
c) After about half an hour it looked like diagram B.
d) Explain why the bag changed shape in this way.
e) What would happen if the bag was placed in a stronger sugar solution than the solution inside it?

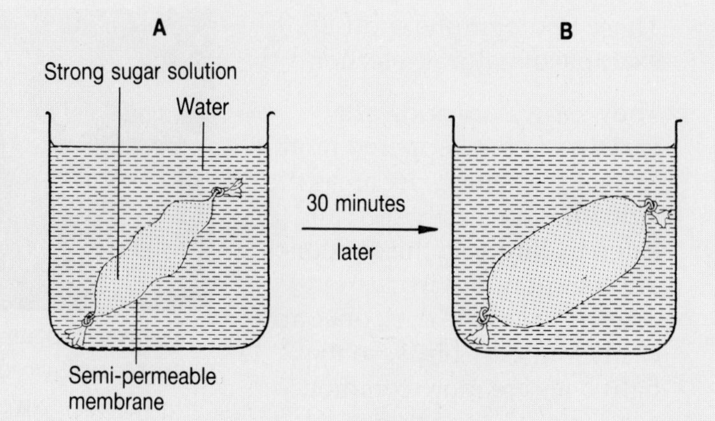

3 An uncooked potato was peeled. A piece from the middle was cut into two strips X and Y, each 50 mm long. One strip was placed in a strong sugar solution. Another strip was placed in pure water.

After 30 minutes strip X had grown longer. Strip Y had shrunk.
a) What could have happened to the cells in each strip, to make one strip swell and the other shrink?
b) Which strip had been placed in the sugar solution? Explain your answer.
c) The experiment was repeated using strips of cooked potato. These strips did not change in length. Can you explain why?

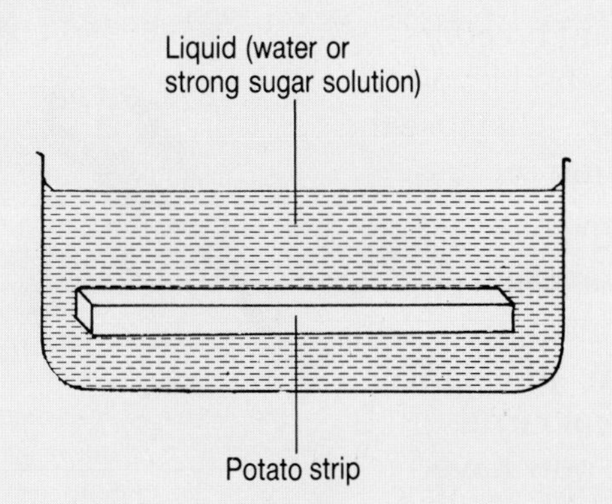

4 **a)** What is a tissue? Name two examples.
b) What is an organ? Name two examples.
c) What is an organ system? Name the organ system which:
carries 'messages' around the body,
changes food into a liquid,
carries food, oxygen and wastes around the body.

5 Some characteristics of living things are affected by their surroundings or environment.

a) Write down the two headings *Affected by the environment* and *Not affected by the environment*.

b) Now sort the list below into two groups under these headings.

human eye colour
a cow's milk yield
size of plant seeds
size of potatoes
number of petals on a flower
human ear shape
human height
a cow's coat colour

6 a) E is the gene for brown eye colour, and e is the gene for blue eye colour. Which gene is dominant?

b) Both father and mother have the genes Ee in their cells. What colour are their eyes?

c) Copy the diagram below. Complete it to show the genes in each sex cell, and how the genes come together at fertilization.

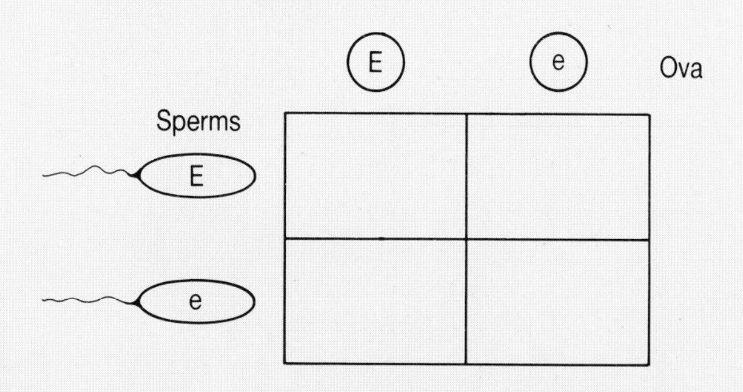

d) Which of the fertilized cells (zygotes) are homozygous for eye colour? Which are heterozygous?

e) Which zygotes will produce children with blue eyes? Which will produce children with brown eyes?

Investigations

1 **Looking at plant cells**

Pull a leaf off a moss plant. Put it on a microscope slide in a drop of water, and place a cover slip on top of it. Look at the leaf under low magnification and you will see how the plant cells are arranged in it. The green blobs in the cells are chloroplasts. Look at a cell under higher magnification to see the chloroplasts more clearly.

2 **Diffusion**

a) Close all the windows and doors of the classroom.

b) Spray some perfume or air freshener into the air at one end of the classroom. Note the time.

c) Ask the members of the class to raise their hands when the smell reaches them.

d) How long does it take for the smell to get everywhere in the room?

e) Explain how the smell spreads through the room.

3 **Variation**

Before starting this investigation, make sure you understand the difference between continuous variation and discontinuous variation (page 31).

a) Ask everyone in your class to measure one of their little fingers to the nearest millimetre, as shown in the diagram opposite.

b) Collect all the measurements. If your class has less than 20 pupils, collect the measurements from another class too. The more results you have, the better.

c) Next copy the graph called a *histogram* started below. Use your measurements to complete it.

d) Is variation in finger length an example of continuous variation or discontinuous variation? Use your histogram to explain your answer.

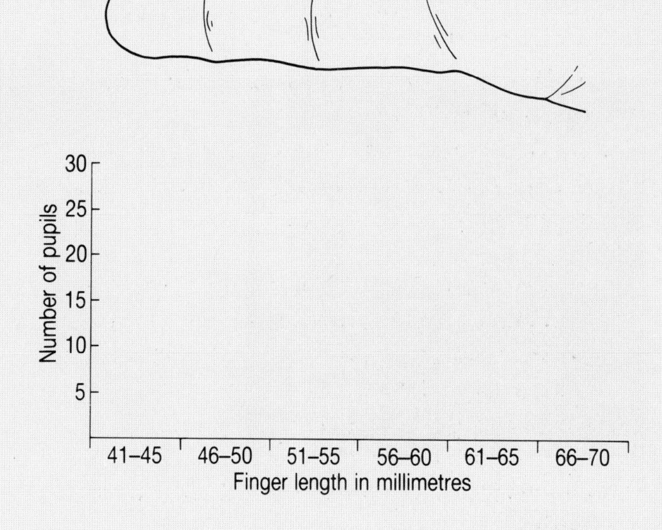

3·1 Flowering plants

There are thousands of plants that flower. Daisies, tomato plants, grasses, and chestnut trees are just some of them. Can you think of others? Flowers have male and female **sex organs**. These make **seeds** from which new plants grow.

The parts of a flowering plant

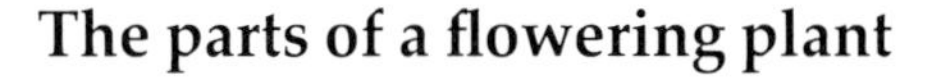

The **flower** contains the male and female sex organs. These make seeds.

The **leaves** make food for the plant by photosynthesis. For this they need sunlight, carbon dioxide from the air, and water and minerals from the soil.

Roots anchor a plant in the soil. They also take in water and minerals from it.

Most of the water and minerals are taken in through the **root hairs**.

Buds contain small, partly-grown, leaves or flowers. The buds protect these young parts.

The **stem** has tubes inside it. Some tubes carry water and minerals from the roots to the leaves. Other tubes carry food up and down the plant.

Soil level

This is a **main root**.

This is a **side root**, or **lateral root**.

Which flowering plant is this?

And can you name this one?

How plants make food

Plants make food by **photosynthesis**. Photosynthesis means making things with light. That is what plants do. They use light energy to make food from carbon dioxide and water.

Inside a leaf is a green substance called **chlorophyll**. It can trap energy from sunlight.

Carbon dioxide is taken into the leaf through tiny holes. Water is carried into the leaf from the stem through tubes.

Trapped light energy makes the carbon dioxide and water combine, to make a sugar called glucose, and oxygen. Glucose is used for food. Oxygen is given off into the air and keeps air breathable for all living things.

At night, photosynthesis stops.

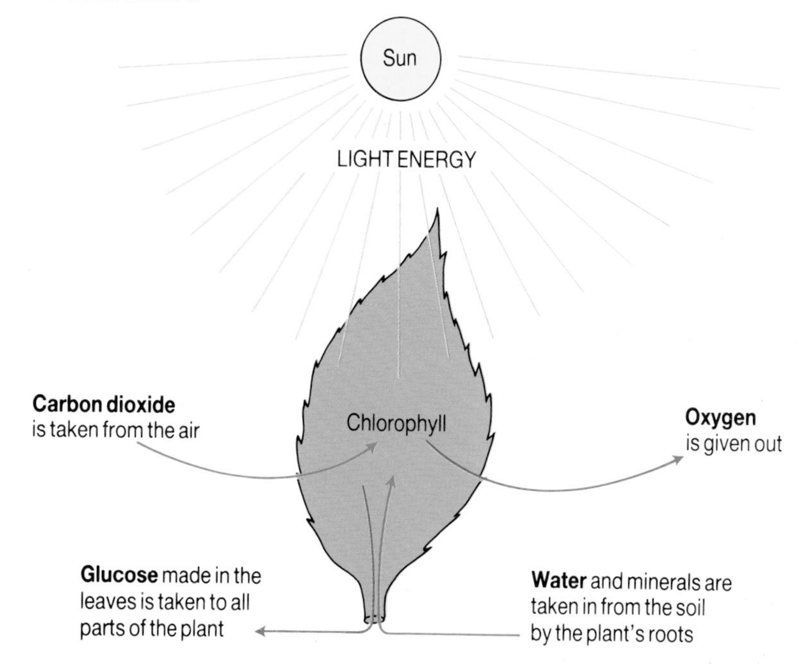

Here is a short way to show what happens during photosynthesis:

$$\text{water} + \text{carbon dioxide} \xrightarrow[\text{chlorophyll}]{\text{sunlight}} \text{glucose} + \text{oxygen}$$

Food factories – busy making food for a horsechestnut tree.

How plants use the glucose from photosynthesis

1 Some is used straight away, to give the plant energy.
2 Some is stored up. It is changed into starch or oil, and stored in stems, roots, seeds, and fruits. When it is needed it is changed back to glucose.
3 Some is used to make cellulose for cell walls.
4 Some is combined with minerals, and used to make proteins and the other things plants need for growth.

Questions

1. Where are the sex organs in a flowering plant?
2. What two things do roots do?
3. What is photosynthesis and where does it occur?
4. What is the green substance in leaves called?
5. What does this green substance do?
6. What does a plant need to make glucose?
7. What does a plant use the glucose for?
8. How is the glucose stored?
9. What else is produced during photosynthesis?

3·2 Leaves

Most leaves are thin and flat, like this:

These are **veins**.

This is the **mid-rib** of the leaf.

The stalk of a leaf is called a **petiole**.

The mid-rib and veins in this leaf are made of tiny tubes. Some of these tubes carry water and minerals into the leaf. Some carry food out to all parts of the plant.

Inside a leaf

This is what a slice of leaf looks like under a microscope.

The **upper skin** of the leaf. It has a layer of wax on it called a **cuticle**. This makes the leaf waterproof.

This thick layer of cells in the middle of a leaf makes glucose by **photosynthesis**.

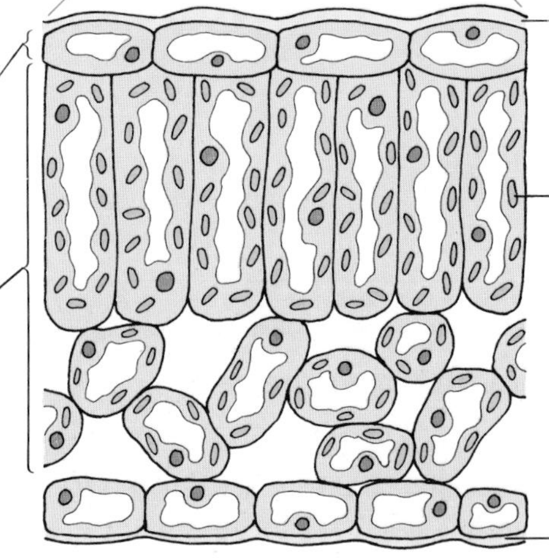

Chloroplasts. These are tiny discs inside cells that contain **chlorophyll**. They trap light energy for photosynthesis.

This is another view of a piece of leaf.

The tiny tubes in a **vein**. Some carry water and minerals into the leaf. Some carry food out.

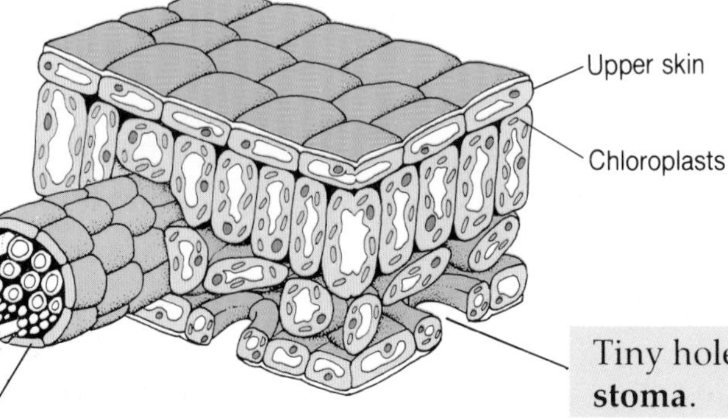

Tiny hole in a leaf called a **stoma**.

Stomata

In most leaves, the lower skin has tiny holes in it called **stomata**. (One hole is called a **stoma**.) Stomata allow carbon dioxide to pass into the leaf, and oxygen and water vapour to pass out.

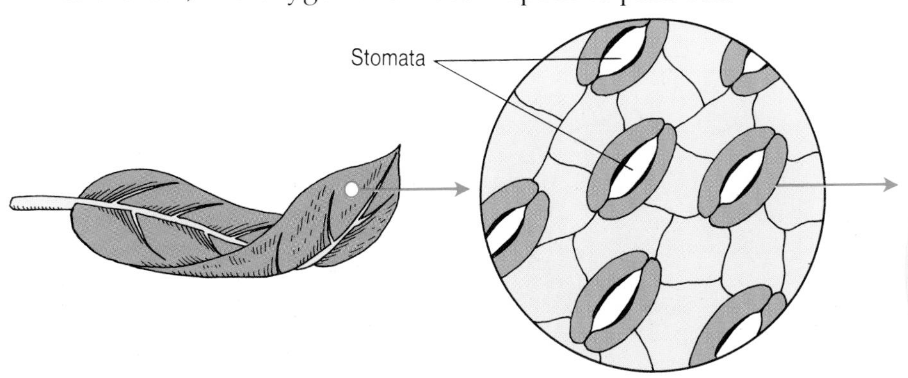

This is what the lower skin of a leaf looks like under a microscope.

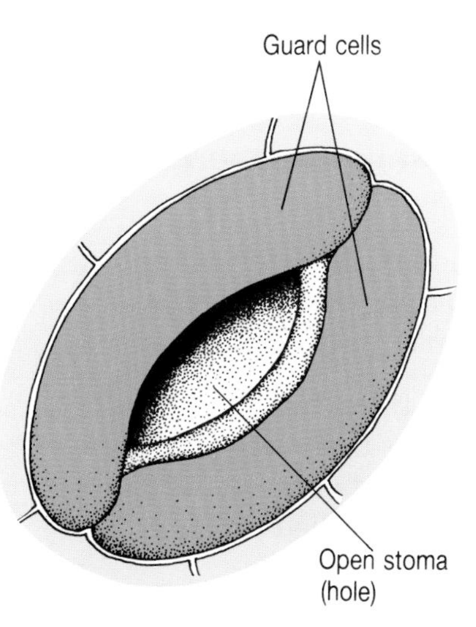

This is one stoma, greatly magnified. It is opened and closed by the guard cells (described in unit 3.4).

How leaves are suited to photosynthesis

1 Many leaves are broad and flat, to absorb as much light as they can.
2 They are thin so carbon dioxide can reach inner cells easily.
3 They have plenty of stomata in the lower skin, to let carbon dioxide in, and oxygen and water vapour out.
4 They have plenty of veins to carry water to the photosynthesizing cells, and carry glucose away.

Dicotyledons, like this rhubarb, have broad leaves with a network of veins.

Monocotyledons, like this grass, have long narrow leaves with parallel veins.

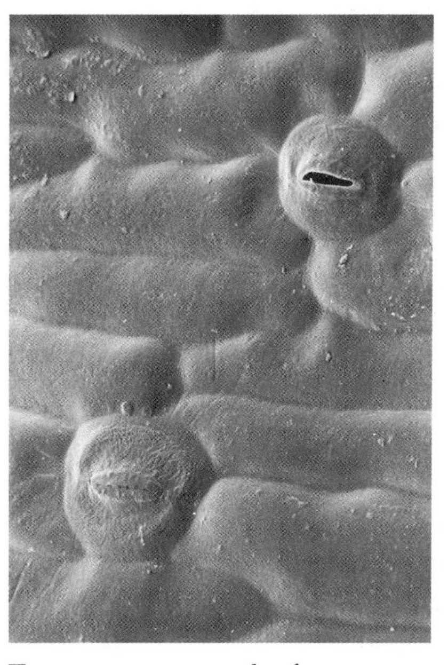

Two stomata on a leaf, magnified by 8000. The top one is open, the bottom one closed.

Questions

1. What makes a leaf waterproof?
2. What are chloroplasts? Where are they found?
3. What is inside a leaf vein?
4. What are stomata? How do they help leaves make food?
5. Explain why leaves are thin and flat, and well-supplied with veins.

3·3 Transport and support in plants

A plant's transport system

A plant has lots of thin tubes inside it. They carry liquids up and down the plant. They are the plant's **transport system**.

Some tubes carry glucose solution from the leaves to every part of the plant. These tubes are called **phloem**.

Some tubes carry water and minerals up from the soil. They are called **xylem vessels**.

Xylem and phloem are grouped together in **vascular bundles**.

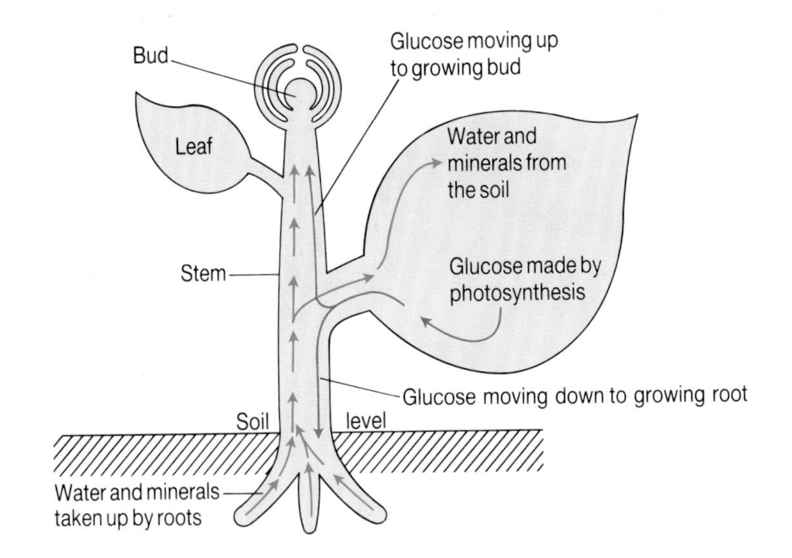

Inside a stem

This is what a thin slice of stem would look like, greatly magnified.

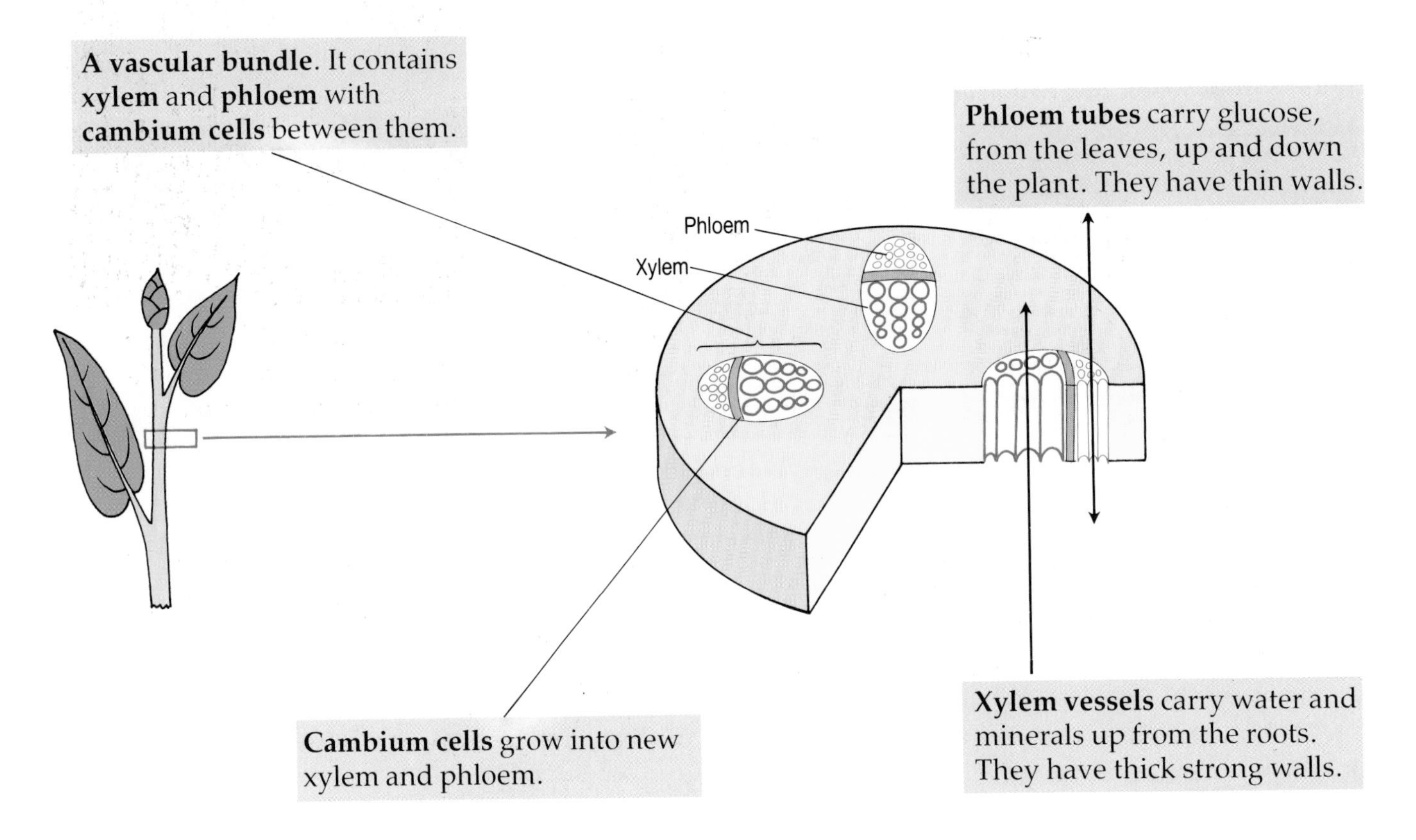

Inside a root

This is what the tip of a root would look like, sliced open and greatly magnified.

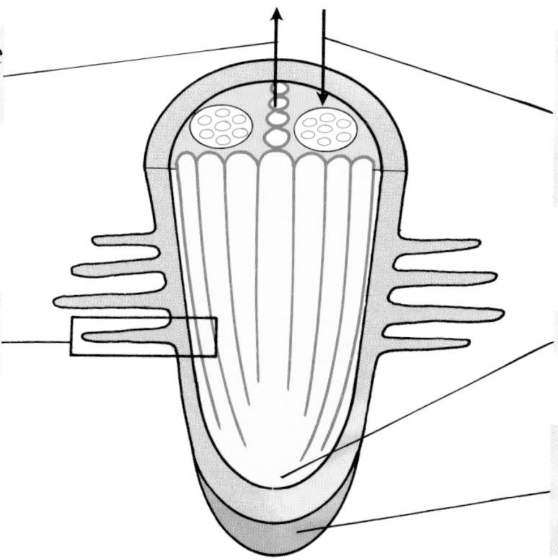

Water and minerals flow up the **xylem vessels** to the stem.

Glucose from the leaves flows down the **phloem tubes**. It feeds the growing cells at the root tip.

Root hairs take in water and minerals from the soil.

Growing cells at the root tip.

The growing tip of the root is protected by a layer of cells called the **root cap**.

Support in plants

Xylem vessels have thick strong walls. So they help to hold up or support a plant. Bushes and trees have plenty of xylem to support them. In fact, tree trunks are mostly made of xylem.

The soft parts of plants, like leaves and flowers, are supported mainly by the water in their cells.

A plant with plenty of water has firm or **turgid** cells. Its stem is straight and its leaves and petals are firm.

A plant without water has soft or **flaccid** cells. Its stem, leaves and petals are soft. They droop, or **wilt.**

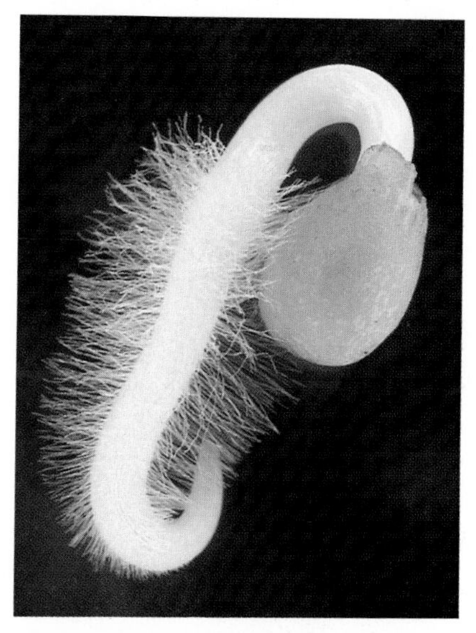

A white mustard seedling. It takes in water and minerals through its tiny root hairs.

Questions

1 **a)** Which tubes in a plant carry glucose?
  b) Where do they carry the glucose from?
  c) Where do they carry it to?

2 **a)** Which tubes carry water and minerals?
  b) Where do they carry them from?
  c) Where do they carry them to?

3 What does a vascular bundle contain?

4 Which cells grow into xylem and phloem?

5 Xylem vessels help support a plant. Why are they able to do this?

6 What is:
  a) a turgid cell?
  b) a flaccid cell?
  c) a root hair?

3·4 Transpiration

A plant loses water vapour through its leaves. This loss of water is called **transpiration**. The water first evaporates from cells inside a leaf. It then escapes through tiny holes in the leaf called stomata (one hole is called a stoma). In most plants stomata are on the undersides of the leaves.

The transpiration stream

Water is lost from the plant's leaves. At the same time more water flows up xylem vessels from the roots to replace it. This flow of water from roots to leaves is called the **transpiration stream**.

The transpiration stream flows fastest on warm dry sunny days, because this is when water evaporates fastest from the leaves. It slows down on cold, dull, damp, days and when a plant is short of water.

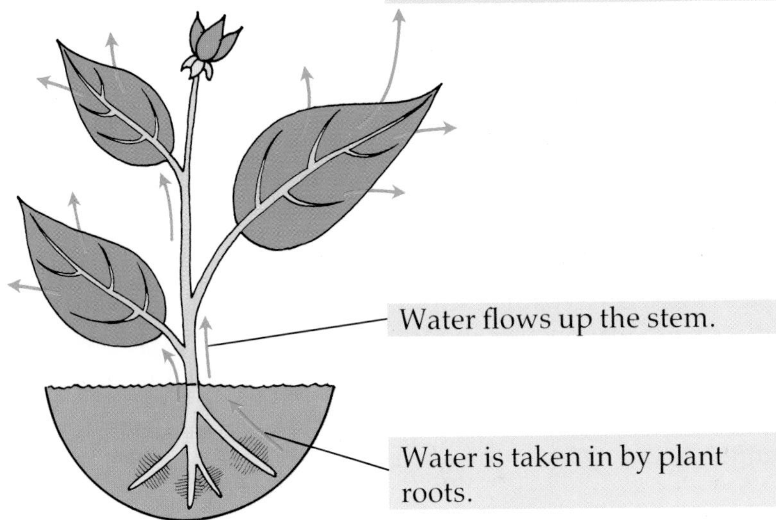

How stoma work

A stoma is a hole between a pair of **guard cells**.

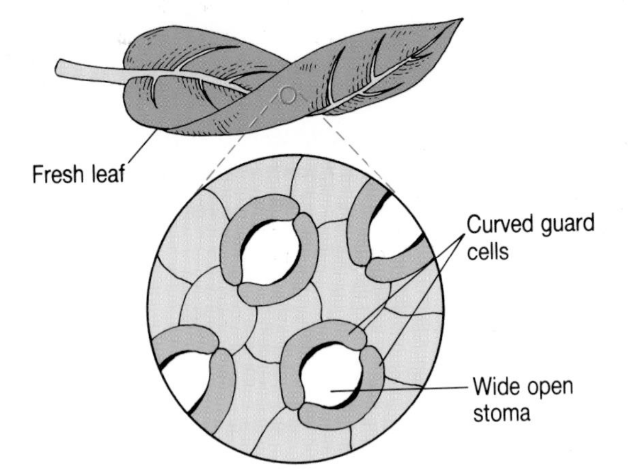

When a plant has plenty of water, the guard cells become curved and the stoma between them opens. This allows water to escape from the leaves of the plant.

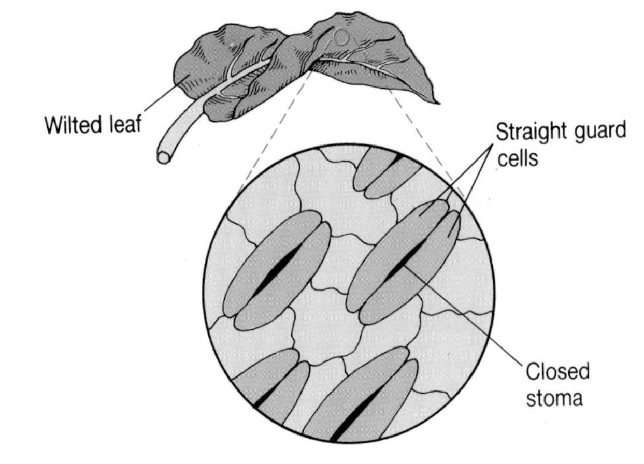

When a plant begins to lose water faster than its roots can take it up, guard cells become less curved. This closes the stomata and slows down the loss of water from the plant.

Why transpiration is important

1 The transpiration stream carries water and minerals from the soil to the leaves. Water is needed for photosynthesis. Minerals are needed for making proteins.

2 Water is also needed to keep cells turgid (firm), so that they support the plant (unit 3.3).

3 The evaporation of water from the leaves keeps them cool in hot weather.

How the transpiration stream flows

1 Root hairs take in water from the soil by osmosis (unit 2.3).

2 Water passes into the root. It moves from cell to cell until it reaches the xylem vessels.

3 Water is sucked up the vessels (like lemonade sucked up a drinking straw). Water passes into the leaves through leaf veins. Water evaporates from leaf cells and escapes through stomata.

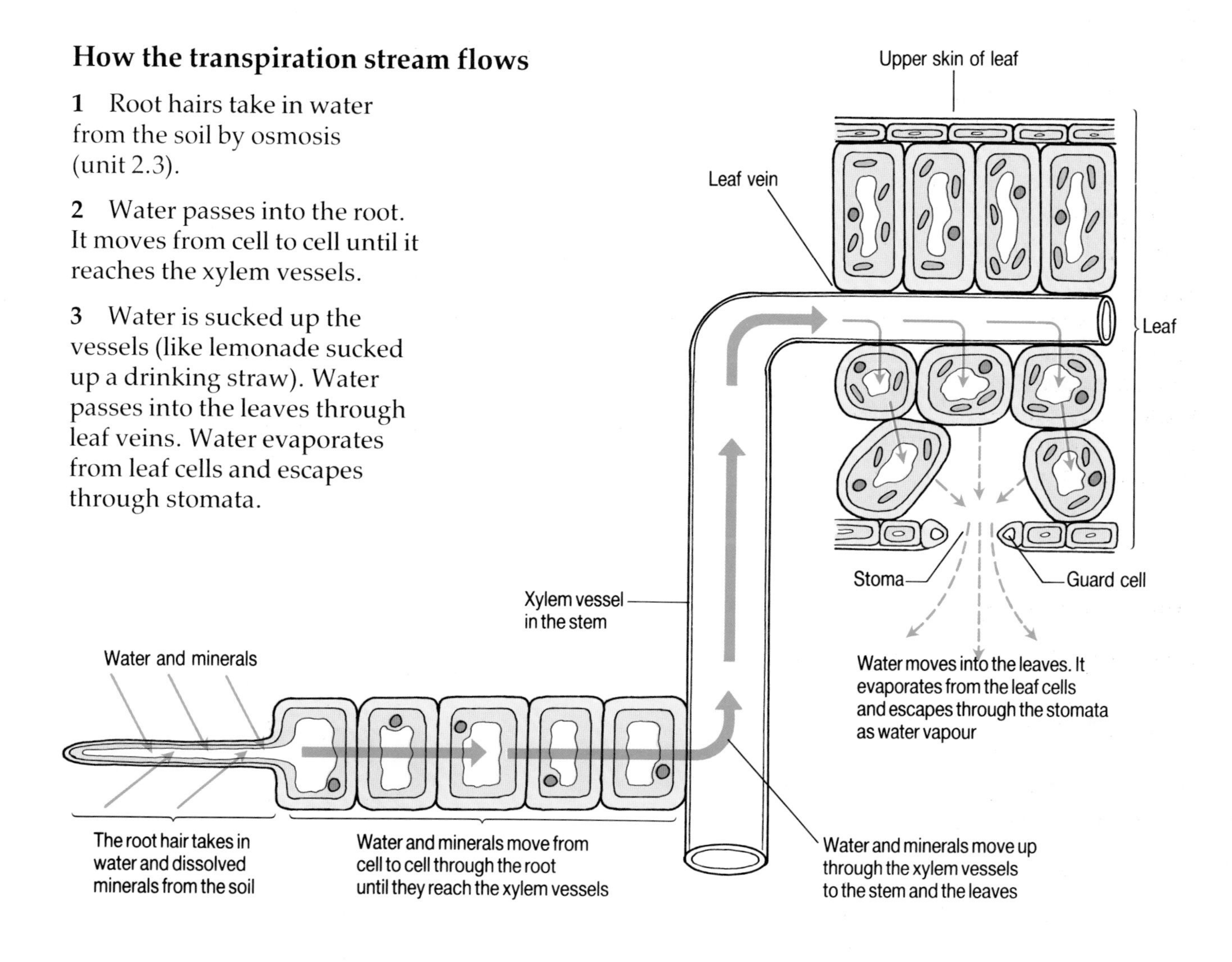

Questions

1 Does all the water which rises up a plant escape through the leaves?

2 a) What is the transpiration stream?
  b) Give three reasons why it is important.

3 When is the transpiration stream fastest?

4 What happens if water is lost from the leaves faster than it is taken up through the roots?

5 Explain how guard cells work.

3·5 Flowers

Humans have male and female sex organs, and so do flowering plants. A plant's sex organs are in its flowers. If you slice a flower open you will see something like this:

This is a male sex organ. It is called a **stamen**.

This is a swelling called a **nectary**. It makes **nectar**, a sweet liquid that attracts insects.

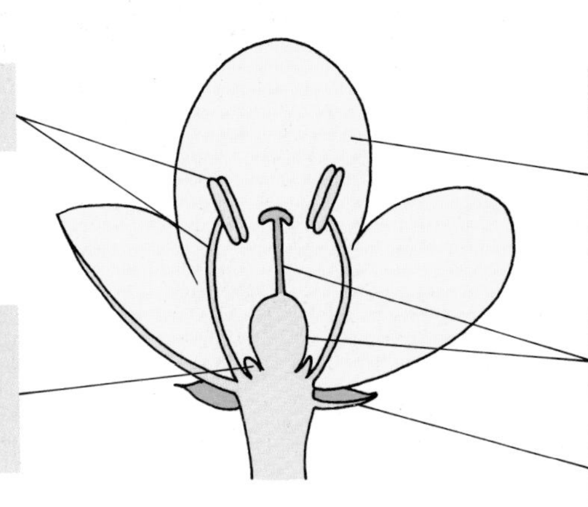

This is a **petal.** Some petals are coloured and scented to attract insects. Insects help plants to make seeds, as you will see later.

This is a female sex organ. It is called a **carpel**.

This is a **sepal**. Sepals protect the flower when it is in bud.

Parts of a stamen

The **anther**. It is made up of four **pollen sacs**. These are full of **pollen grains**. Pollen grains contain the male sex cells of the plant.

The stalk or **filament** of the stamen.

Anther cut open.

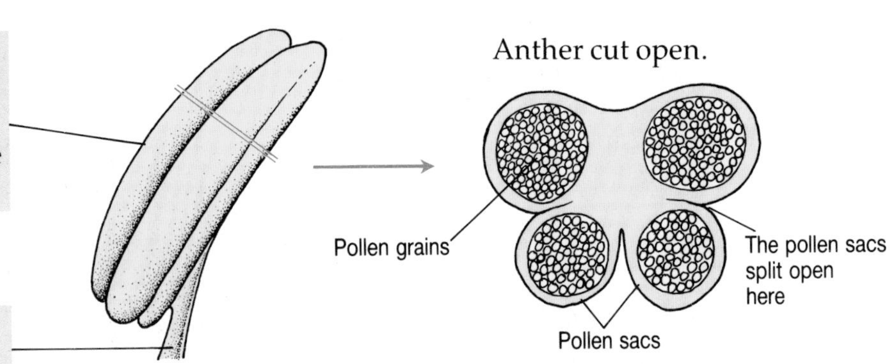

Parts of a carpel

The **stigma**. Pollen grains stick to this during pollination.

The **style**.

The **ovule** contains the female sex cell. It forms a seed after fertilization.

The **ovary** protects the ovule.

The **female sex cell**.

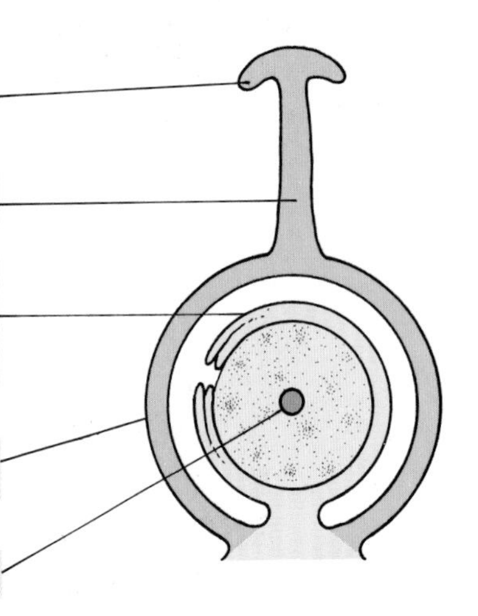

A tulip carpel surrounded by brown stamens. Note the large stigma at the top of the carpel. Why does it need to be sticky?

Some different flowers

There are many different kinds and shapes of flowers. They have different numbers of petals, stamens, and carpels. Some flowers, such as grasses, do not have any petals or nectaries.

The petals on a plum flower are all the same shape. Note the stamens sticking out. The stamens have long filaments with the anthers at the end.

A grass flower doesn't have petals, scent, or nectar. You can see it has very large anthers.

Sunflowers are made up of hundreds of small flowers, called florets, gathered together.

The petals on a pea-flower are of different shapes and sizes.

Questions

1. Why are some petals coloured and scented?
2. What is the name of:
   a) the male sex organ of a flowering plant?
   b) the female sex organ of a flowering plant?
3. What do sepals do?
4. What is nectar? Where is it made?
5. Copy and complete:
   The _____ is made up of four pollen _____ .
   These are full of _________ .
6. What is inside a pollen grain?
7. Draw a carpel and label the parts.
8. Draw a stamen and label the parts.

3·6 How fruits and seeds begin

A new seed begins when the male sex cell in a **pollen grain** joins up with a female sex cell in an **ovule**. The first step in making this happen is called **pollination**. Pollination takes place when pollen is carried from an anther to a stigma.

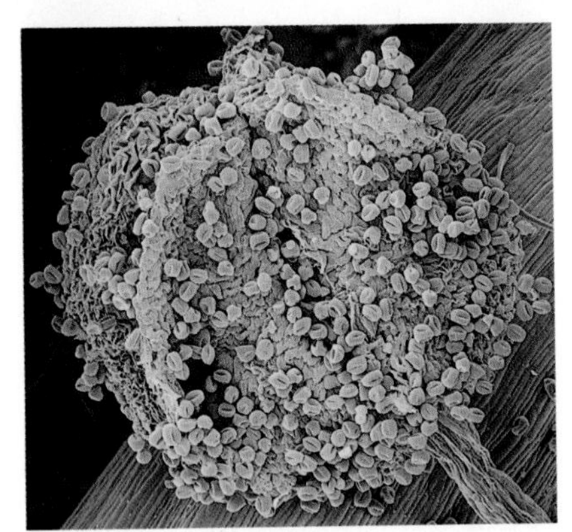

First the anther ripens. The pollen sacs split open releasing pollen grains.

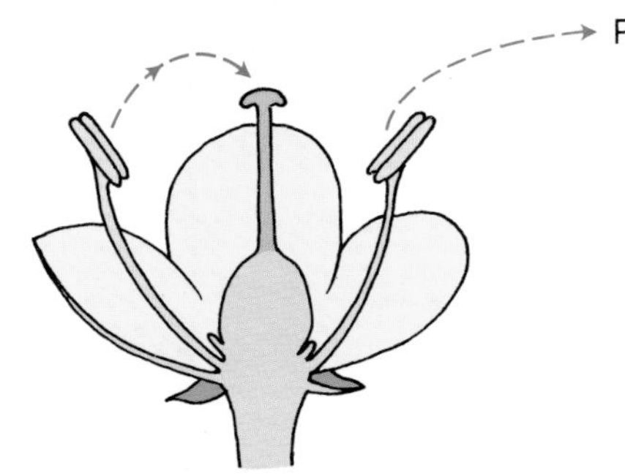

Pollen may be carried to a stigma in the *same* flower. This is called **self-pollination**.

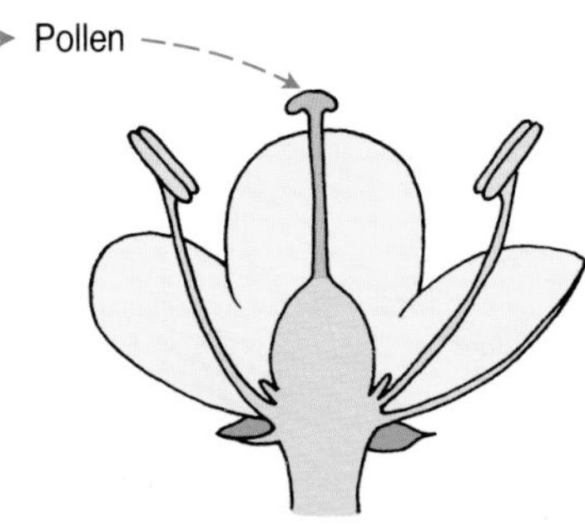

Or pollen may be carried to a stigma in another flower. This is called **cross-pollination**.

Pollination by insects

Bees and other insects can carry pollen from flower to flower. That is what happens with buttercups, dandelions, and foxgloves.

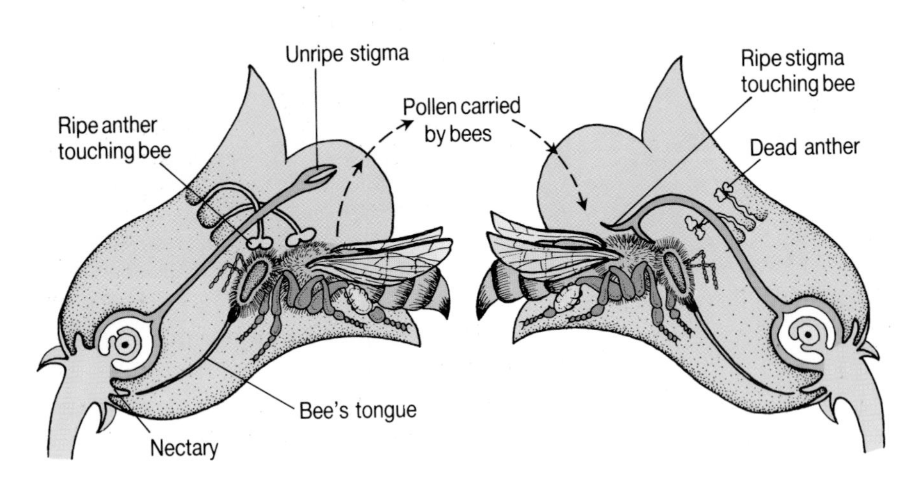

Insect-pollinated flowers have:

1 large, coloured, scented petals and nectar, to attract insects
2 large pollen grains that can stick to an insect's body
3 anthers and stigmas inside the flower, so that the insect can brush against them when it is drinking nectar.

A bee with foxglove pollen clinging to its body. Inside this foxglove, the pollen will rub off on the sticky stigma.

Pollination by wind

The wind can carry pollen from flower to flower. That is what happens with grass, stinging nettle, and catkins.

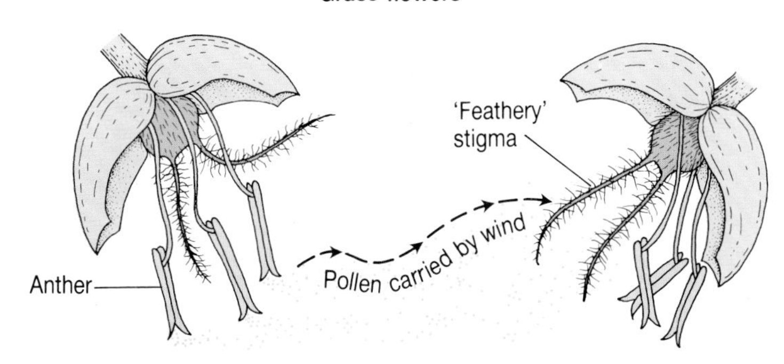

In catkins, the male and female sex organs are in separate flowers. Wind shakes the yellow pollen from the male catkin. . .

. . .and carries it to the female catkin's sticky stigmas.

Wind-pollinated flowers do not have large scented petals, or nectar, because they do not need to attract insects.

They do have:

1 anthers that hang outside the flower, to catch the wind
2 large amounts of light, small pollen grains that blow in the wind
3 spreading, feathery stigmas to catch air-borne pollen grains.

Fertilization

After a pollen grain lands on a stigma the next step is **fertilization**. Fertilization happens when a male sex cell joins up with a female sex cell.

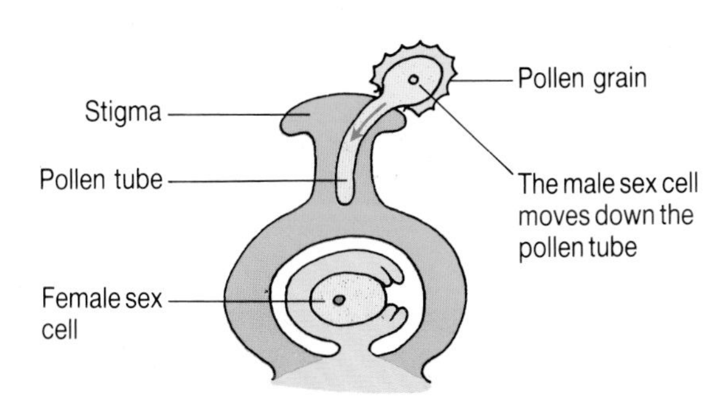

1 First a tube grows out of the pollen grain. It grows towards the female sex cell. A male sex cell moves down the tube.

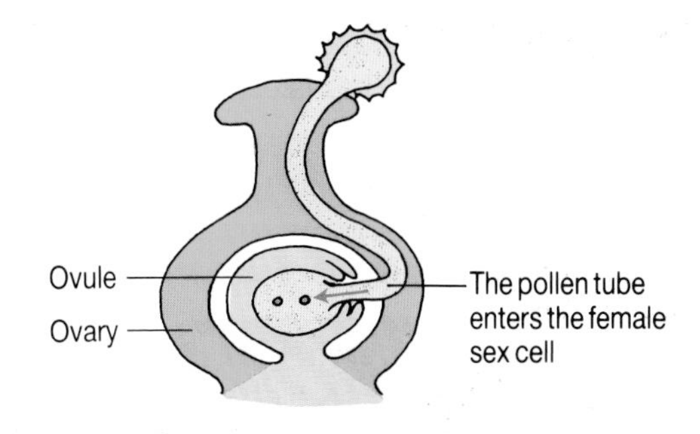

2 The tube enters the female sex cell. The tip of the tube bursts open. The male sex cell joins up with the female sex cell. The ovule then becomes a **seed**. The ovary becomes a **fruit** with the seed inside it. The petals die and drop off.

Questions

1 What are self-pollination and cross-pollination?

2 a) Describe the differences between insect-pollinated, and wind-pollinated flowers.

b) Give the reasons for these differences.

3 What happens after a pollen grain lands on a stigma?

3·7 More about fruits and seeds

In biology a **fruit** is any part of a plant that contains **seeds**.

Apples, oranges, lemons and tomatoes are all fruits. Think of the seeds inside them.

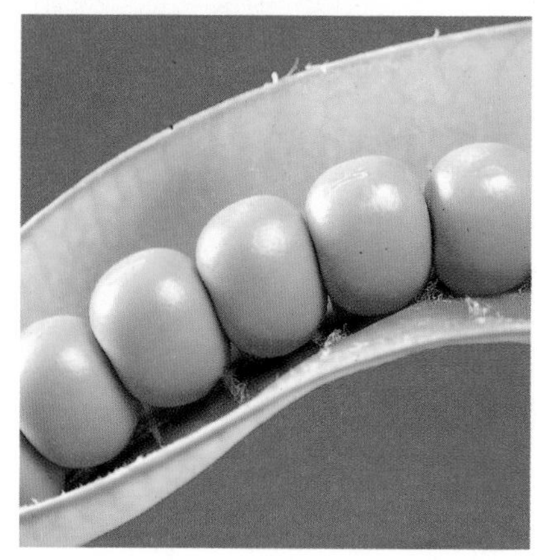

Pea pods are also fruits. The peas inside them are seeds.

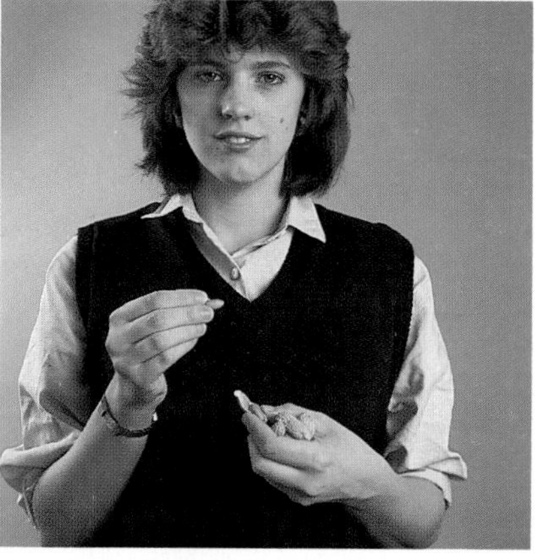

Nuts are fruits too. The part we eat is the seed.

How fruits and seeds get scattered

Most plants produce hundreds of seeds at a time. If they all fell close to the parent plant they would be too crowded. The new plants would have to compete with each other for water, minerals, and light.

Instead, fruits and seeds get scattered by the wind, or animals, or by the plant itself. This is called **dispersal** of fruits and seeds.

Dispersal by wind

These fruits and seeds are dispersed by the wind.

Dandelion and thistle seeds have a downy 'parachute' which floats in the wind.

Sycamore, ash, and lime seeds have wings which carry them long distances in the wind.

Poppy heads sway in the wind scattering seeds through holes in their sides.

Dispersal by animals

These fruits and seeds are dispersed by animals.

Blackberry, sloe, and hawthorn berries are eaten whole by animals. The seeds pass unharmed through their bodies.

Burdock has hooks which catch on animals' fur or on our clothes. The seeds can be carried long distances before falling off.

Acorns and beech nuts are carried away by birds and squirrels. Some are dropped before they are eaten.

Dispersal by plants

Some plants throw away their seeds.

Lupin seed pods suddenly burst open when dry. Their sides coil up, scattering the seeds.

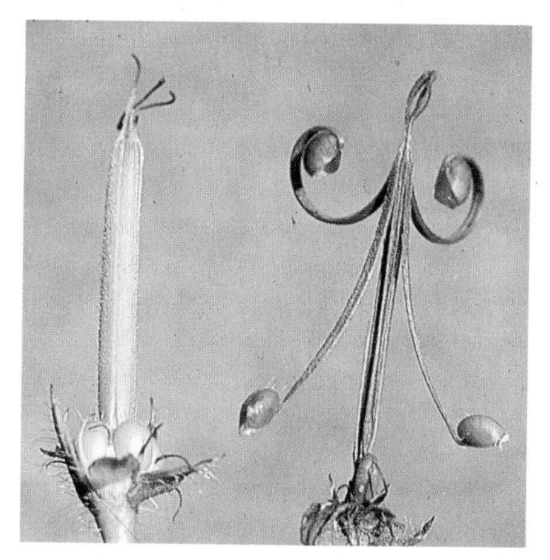

Geranium fruits also burst open when dry. Their sides spring upwards, scattering the seeds.

Some seeds are collected and stored and sold by man. They get dispersed by gardeners and farmers.

Questions

1 In biology, what is a fruit?

2 Why do seeds need to be dispersed?

3 **a)** Name two plants that use wind dispersal.
b) How are their fruits and seeds suited to this?

4 How do some plants disperse their own seeds?

5 Describe two ways in which animals can disperse fruits and seeds. Give an example of each.

6 Describe how you may have helped disperse seeds without knowing it.

3·8 How seeds grow into plants

A seed looks dead. But inside it there is a tiny plant called an **embryo**, and a store of food. If the seed is given water, air, and warmth it 'comes to life', and the embryo begins to grow. This is called **germination**.

The embryo in a seed has three parts. It has a root or **radicle,** a shoot or **plumule,** and one or more seed leaves called **cotyledons.** The embryo needs its stored food for growth, until it has enough leaves to make food by photosynthesis.

Not all seeds get a chance to grow into plants. Some are eaten first – like these baked beans.

Bean seeds

Bean seeds have two seed leaves, so they are called **dicotyledons** (di means two). The seed leaves of beans are swollen, because they contain the stored food.

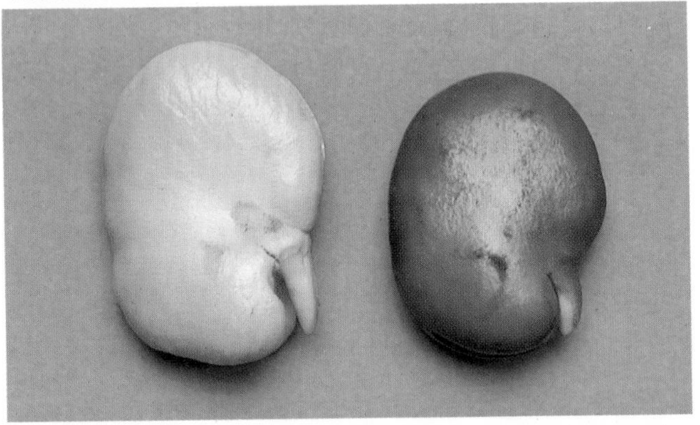

Broad beans. The one on the left has had its seed coat removed.

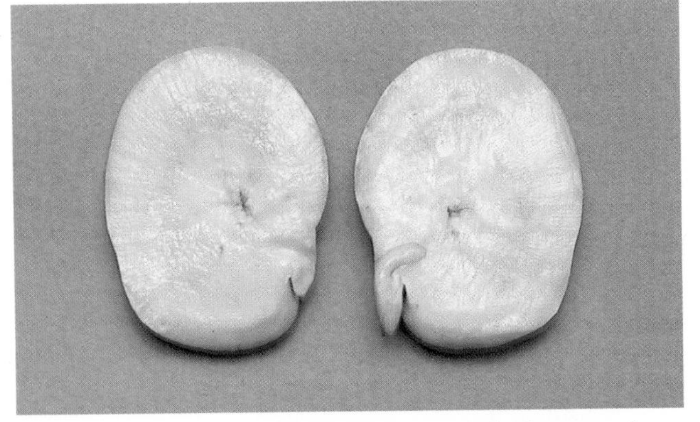

Broad bean cut in half to show cotyledons and embryo.

Germination of a French bean seed

When a French bean germinates, the cotyledons come above the ground. This is called **epigeal germination**. These pictures show a French bean germinating. The brown object in two of the pictures is the seed coat which has split off the seed.

The cotyledons push their way out of the soil.

They provide the food the plant needs, until its leaves grow.

The leaves are growing, so the cotyledons will soon drop off.

Maize and other cereal grains

Maize, wheat, rice, barley, and other cereal grains have only one cotyledon. So they are called **monocotyledons** (mono means one). Their food is not stored in the cotyledon, but in a mass of cells called **endosperm** in a space above the cotyledon. When grains are milled, the endosperm spills out. We call it flour.

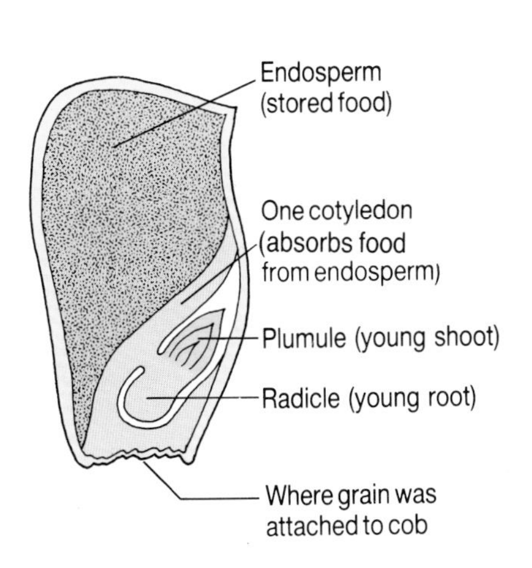

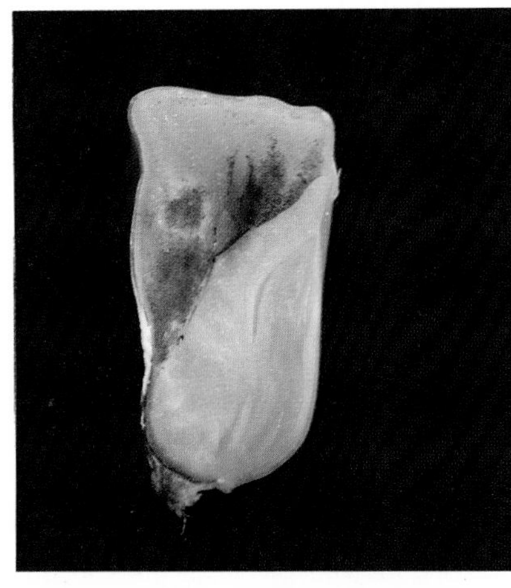

A maize grain cut in half. Compare it with the drawing.

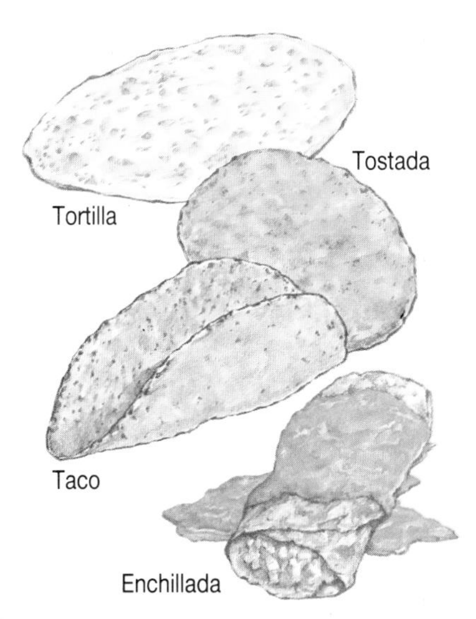

We make a lot of our food by grinding up the seeds of maize, wheat and other cereals, to get their food stores. Above are some Mexican foods made from maize flour.

Germination of maize grains

When a maize grain germinates, the cotyledon stays below the ground. This is called **hypogeal germination**. Other grains, and broad beans, also germinate in this way.

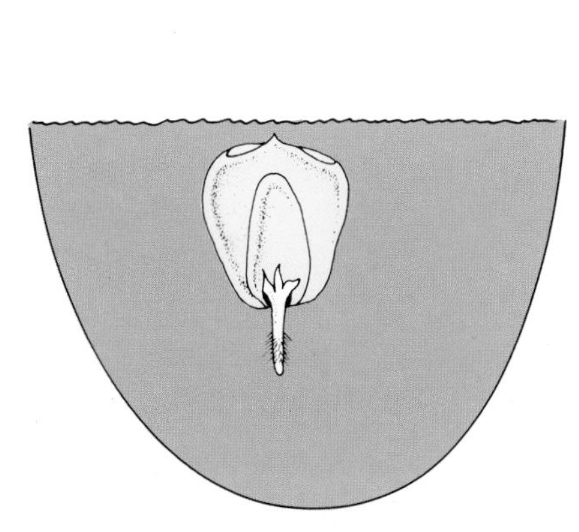

A maize grain germinating. Note the root growing downwards. Its tip is protected by a root cap.

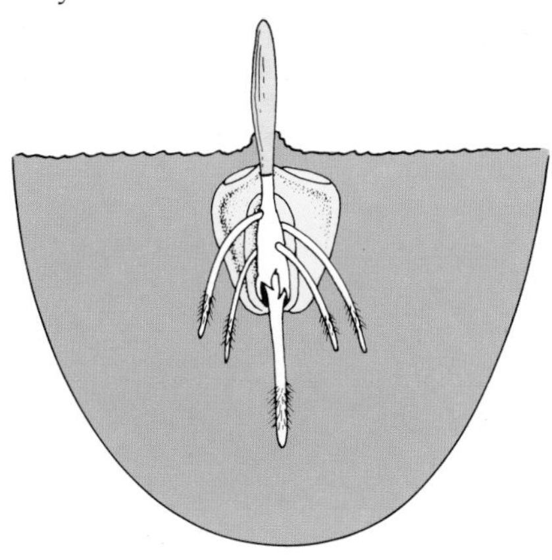

The young shoot pushes through the soil. The endosperm provides the food it needs.

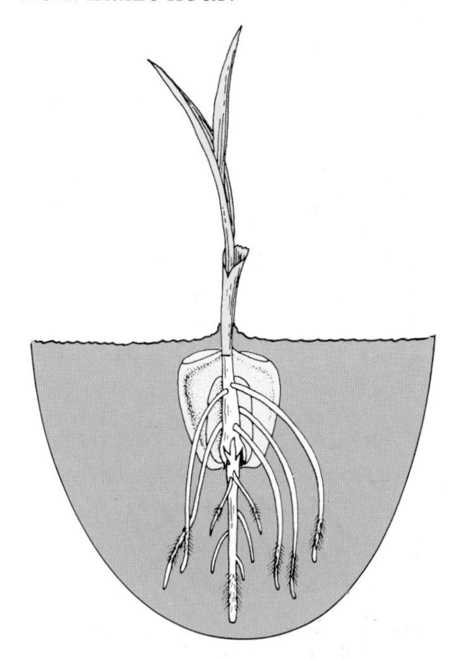

The new leaves can now make their own food. The cotyledon is still below ground.

Questions

1 What three things does a seed need to grow?

2 What is inside a seed?

3 What is:
a) a radicle? **b)** a plumule? **c)** a cotyledon?

4 Why do seeds need a store of food?

5 Is maize grain a dicotyledon? Explain.

6 What is the main difference between epigeal germination and hypogeal germination?

3·9 Plant senses

Plants have no eyes or ears or noses. But they can still sense things. They can sense light, and the pull of gravity, and water. They respond to these things by growing slowly in certain directions. These responses are called **tropisms**.

Phototropism: response to light

Plants need light to make food. So they respond to light by growing towards it. They also turn their leaves to face the light. This makes sure leaves get as much light as possible.

These cress plants were grown with light coming from above.

These were grown with light coming from the right.

Geotropism: response to gravity

No matter how you plant a seed in the earth, it can sense which way is down, and which is up. It can sense this from the pull of gravity.

Roots grow down in response to gravity. This makes sure they find soil and water. Shoots always grow up. This makes sure they reach light.

Roots responding to gravity. Which seed was planted upside down?

Hydrotropism: response to water

Roots always grow towards water, even if this means ignoring the pull of gravity. If they have to they will grow sideways, or even upwards, to reach water.

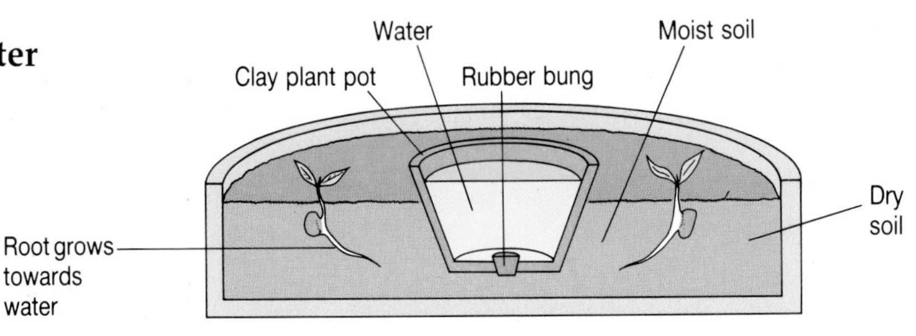

Auxin

A plant's responses to light and gravity are controlled by a chemical called **auxin**. Auxin is made by cells at the tips of stems and roots. It speeds up growth in stems. It slows down growth in roots.

Phototropism and auxin

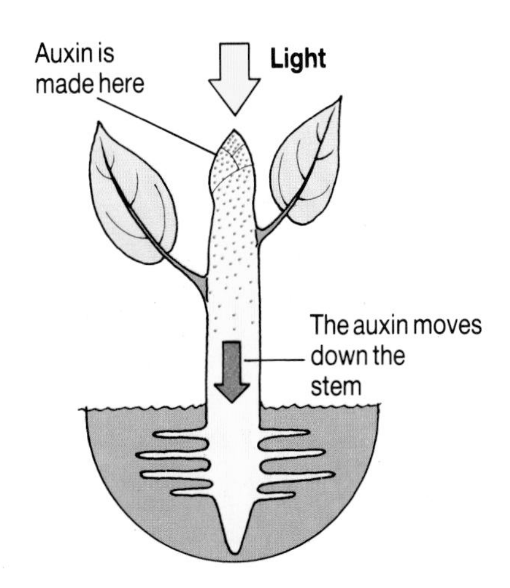

When light comes from above, auxin spreads evenly down the stem. The stem grows straight up.

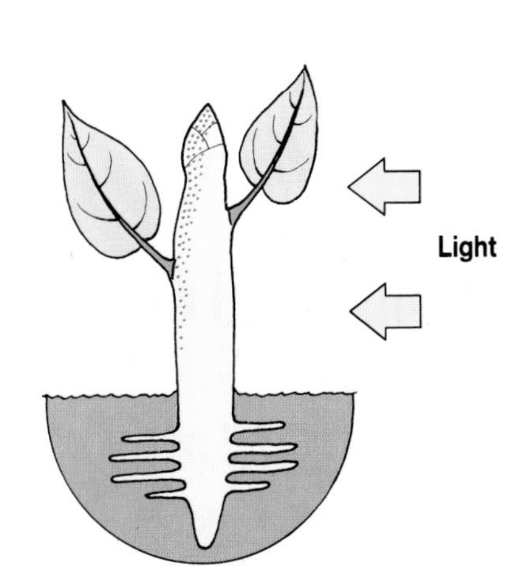

When light comes from one side, auxin spreads down the *shaded* side of the stem.

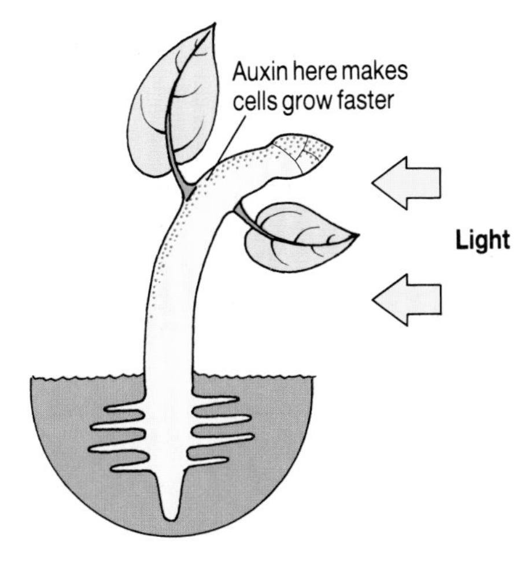

Auxin makes stem cells *grow faster*. This causes the stem to bend towards the light.

Geotropism and auxin

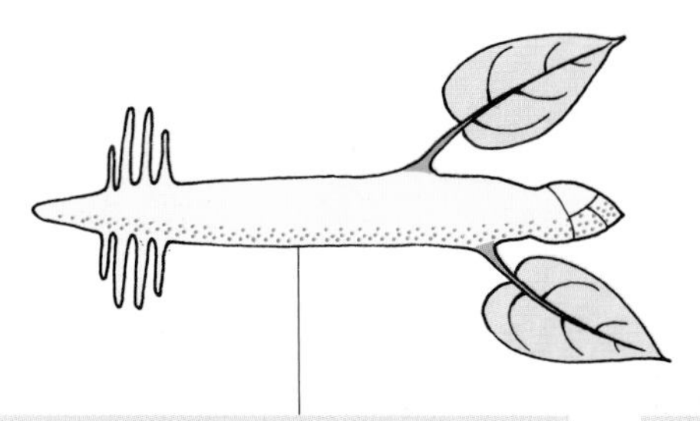

If a plant is laid on its side auxin gathers in the lower half of the stem and root.

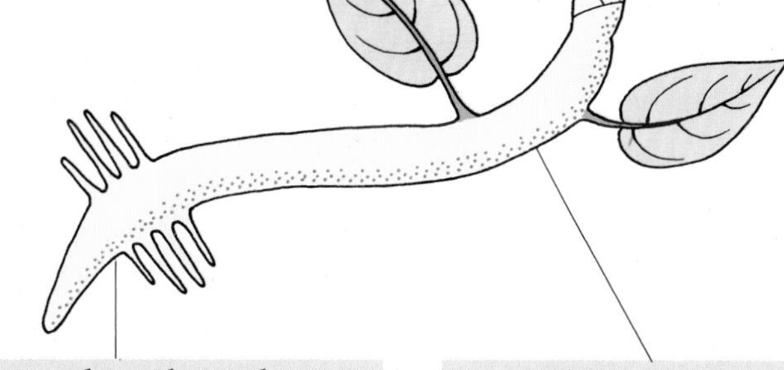

Auxin here *slows down* the growth of root cells. The root curves downwards.

Auxin here *speeds up* the growth of stem cells. The stem curves upwards.

Questions

1 Copy and complete:
   a) A plant's response to light is called ________ .
   b) A plant's response to gravity is called ________ .
   c) A plant's response to water is called ________ .

2 What does a shoot do in response to light, and in response to gravity? How are these responses useful to the plant?

3 Name two responses controlled by auxin.

4 In which parts of a plant is auxin made?

5 How does auxin affect root and shoot cells?

3·10 Asexual reproduction

Sexual reproduction needs *two* parents. They produce the male and female sex cells described in unit 3.5.

Many flowering plants can reproduce in another way, which needs only *one* parent. A plant grows parts which break off, and become new plants. This is called **asexual reproduction** or **vegetative reproduction**. Strawberry, iris, potato and gladiolius plants can reproduce this way. The photos below show how.

Strawberry runners. These are side shoots which grow over the soil. They have buds which grow into new plants. Then the runners die away.

Potato tubers. A potato is really an underground stem of a potato plant, swollen with stored food. The 'eyes' on the potato sprout into new plants.

Iris rhizomes. These are stems that grow sideways through the soil. They are swollen with stored food. In the spring they produce new plants.

Gladiolus corms. A corm is a short stem, swollen with stored food. The corms in the photo will separate from each other and form new plants.

Questions

1. What is the difference between sexual reproduction and asexual reproduction?
2. How do strawberry and potato plants reproduce asexually?
3. What use do plants make of food stored in tubers, corms, and rhizomes?
4. Match these words with the descriptions below: rhizome, tuber, corm, runner
   a) underground horizontal stem,
   b) short stem full of stored food,
   c) its eyes can produce new plants,
   d) grows horizontally above the ground.

3·11 Artificial propagation

You can grow many plants from one plant, by using man-made methods. This is called **artificial propagation**.

Cuttings

A cutting is part of a shoot, root, or leaf cut from a plant. The cutting grows all its missing parts to become a complete plant. Cuttings allow you to produce many new plants from one plant without waiting for flowers and seeds. The new plants are *exact* copies of the parent plant.

A cutting from a plant. See how new roots have grown.

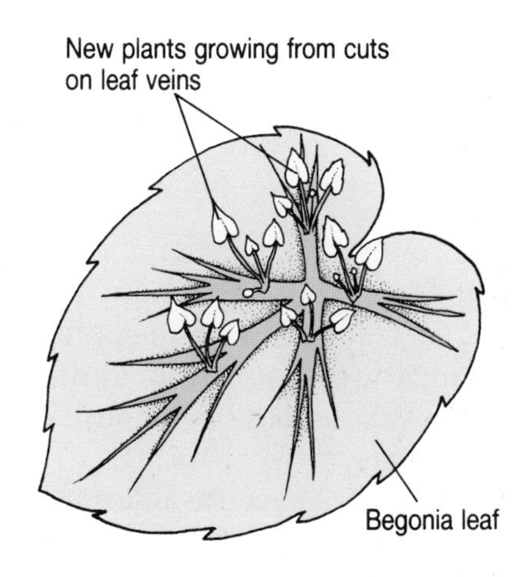

You can grow a begonia plant from a begonia leaf, by making cuts into the leaf veins and putting the leaf in water.

Grafting

You can join part of one plant onto another plant of the same sort. This is called **grafting**. To make a graft, you cut a piece off one plant and join it to a cut on another plant. The cuts soon heal, and the two parts become one plant.

Grafting is used to breed flowering bushes and fruit trees. This is useful because these plants can be very difficult to grow from seeds or cuttings.

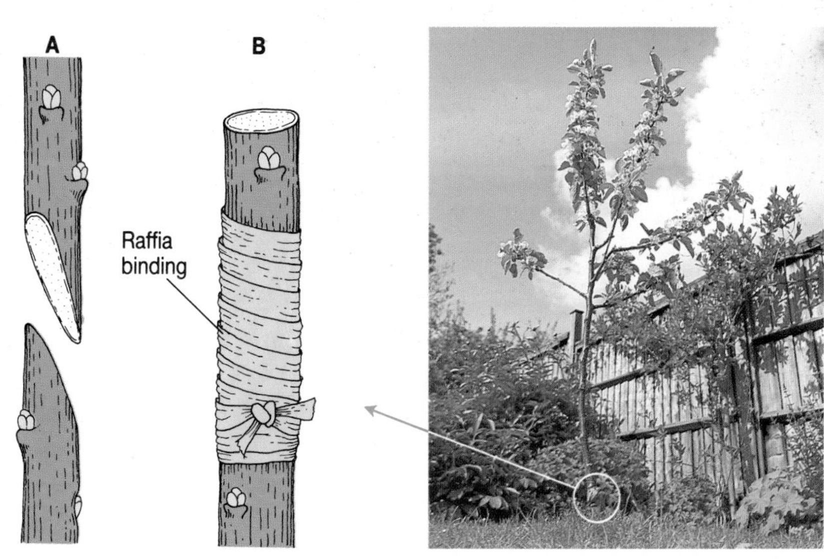

Grafting was used to grow this apple tree.

Questions

1 Why is it useful to be able to grow plants from cuttings and graftings?

2 What is the difference between a cutting and a graft?

Questions

1 The diagram opposite shows the inside of a leaf, highly magnified.

a) Name the parts labelled A to G.

b) For each description below, write down the matching name and label:
cells which make food by photosynthesis;
layer of wax which makes leaf waterproof;
hole which lets gases in and out of leaf;
contains tubes carrying sugar and water;
has chlorophyll inside them.

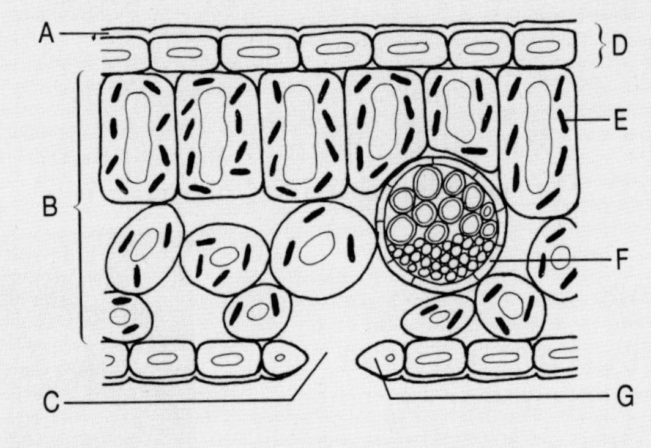

2 A pond contains many water plants. The graphs opposite show how the amounts of oxygen and carbon dioxide in the pond water change during the 24 hours from midnight to midnight.

a) When does the amount of oxygen start to increase?

b) When does the amount of oxygen stop increasing?

c) Why does the amount of oxygen increase during this time?

d) What happens to the amount of carbon dioxide in the pond, while the oxygen is increasing?

e) Why does the amount of carbon dioxide change in this way?

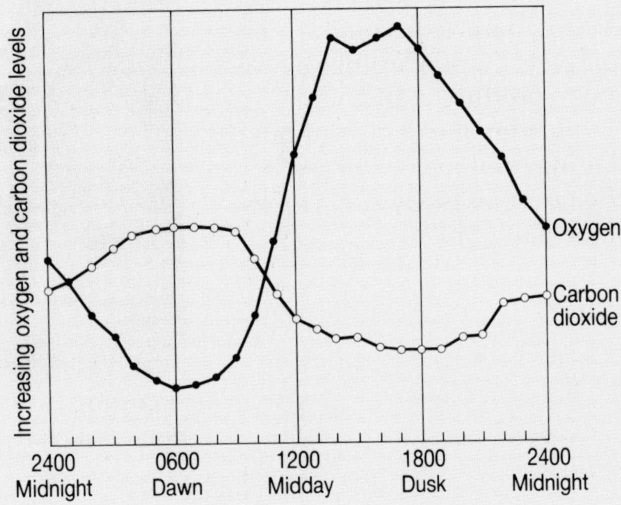

3 The diagrams opposite show a stem and root cut open.

a) Name the parts labelled A to H.

b) For each description below, write down the matching name and label:
tubes which transport water from the roots;
tubes which transport sugar from the leaves;
cells which grow into xylem and phloem;
protects growing cells at the root tip;
absorbs water and minerals from the soil;
cells which produce auxin.

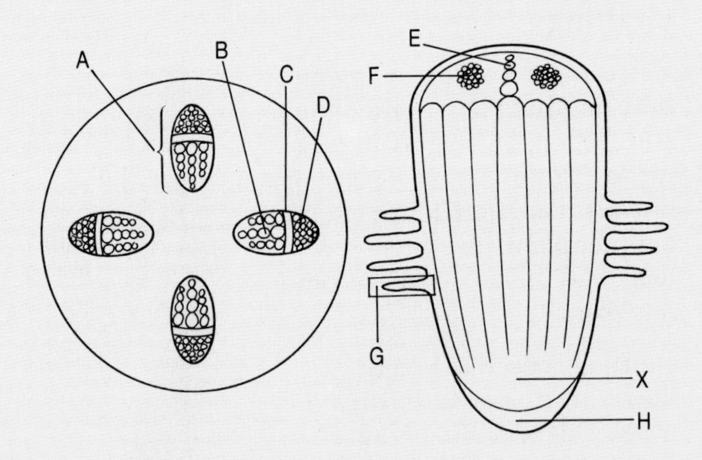

4 The experiment shown opposite used oat seedlings that were prepared in different ways. They are growing in a box with a hole cut at one end.

a) Which seedlings have grown towards the light?

b) Some seedlings have not grown towards the light. Explain why.

c) Explain how these results show that it is the tip of an oat seedling which allows it to bend towards the light.

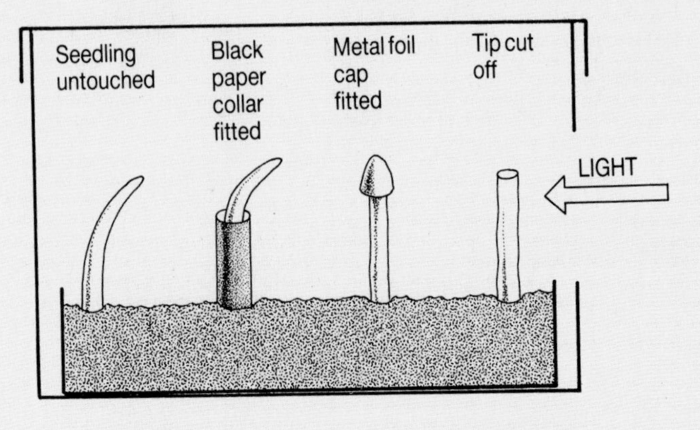

Investigations

1 Photosynthesis

Plants make sugar during photosynthesis. Then they change the sugar into starch and store it for a time in their leaves.
You can show that plants need light and chlorophyll for photosynthesis by using leaves from a *Tradescantia* plant. Its leaves have green and white stripes. Only the green stripes contain chlorophyll.

a) Use a paper clip to fasten a strip of black paper to one leaf (see drawing). Leave the plant in bright sunlight for a day.

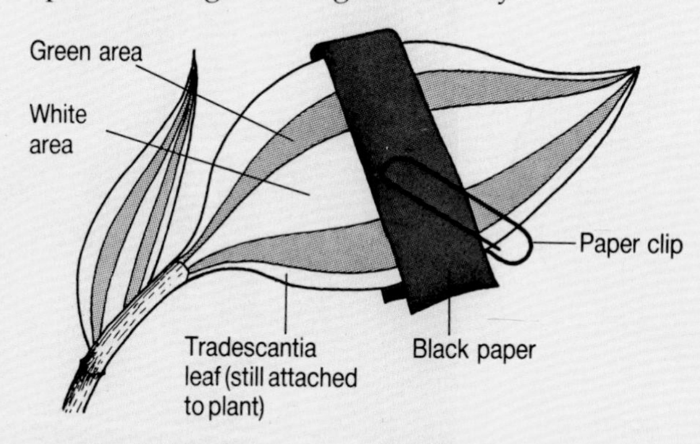

b) Take the paper off the leaf, and pull the leaf off the plant. Boil the leaf in water for 30 seconds.

c) Put the leaf in a test tube of alcohol. Put the test tube into a beaker of boiling water. The alcohol will boil and take all the chlorophyll out of the leaf.

d) Spread the leaf on a white tile and drop iodine on to it. A blue-black colour will appear in the parts of the leaf containing starch.

The leaf will look like this:

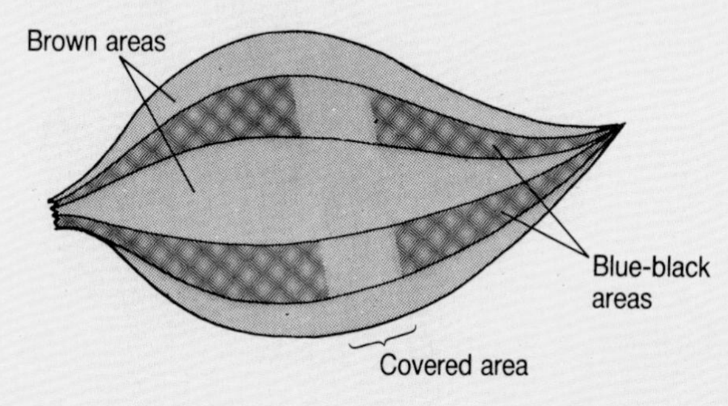

Only the green parts of the leaf which were exposed to light have turned blue black. This shows that plants need light and chlorophyll to make food.

2 Transport in plants

a) Obtain a stick of celery that still has leaves attached.

b) Stand the stick in a beaker of red ink.

c) After about an hour examine the stalk and veins of the leaves. What do you notice? Explain what has happened.

d) Now cut across the stalk. You will see that some parts of it are red. These parts contain xylem vessels. What do xylem vessels do? Cut the stalk lengthwise to see how the xylem vessels run down it.

3 Germination

You can find out what seeds need to germinate (start growing), using mustard or cress seeds and five test tubes.

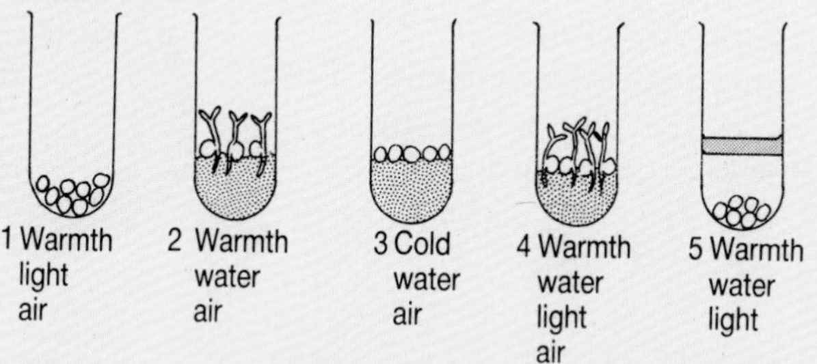

a) Put a few seeds in tube 1. Put the tube in a warm, well-lit place.
These seeds have warmth, air, and light but no water.

b) Put a few seeds on wet cotton wool in tube 2. Leave the tube in a warm dark place. These seeds have warmth, air, and water but no light.

c) Put a few seeds on wet cotton wool in tube 3. Leave the tube in a refrigerator. These seeds have water and air but no warmth or light.

d) Tube 4. Put a few seeds on wet cotton wool in the tube. Leave the tube in a warm, well-lit place. These seeds have warmth, water, air, and light.

e) Put a few seeds in tube 5, and cover them with water which has been boiled and cooled. Then pour a little olive oil onto the water. Put the tube in a warm, well-lit place. These seeds have warmth, water, and light but no air.

f) After three days, examine the seeds in each tube. What do you notice?

g) Now write down all the things that seeds need for germination.

4·1 Skeletons

Jellyfish don't need skeletons since the water supports them. On land they just flop.

Some animals, like the jellyfish in this photograph, can live without a skeleton. They are supported by the water in which they float. But without your skeleton you would collapse in a heap. Like most creatures, especially land animals, you need a skeleton to support your body and give it shape.

Animals with liquid skeletons

Earthworms, caterpillars, and slugs are given shape and support by **liquid** inside them.

This liquid is mainly water. It fills the animal's cells and the spaces inside its body. Muscles squeeze against the liquid. This keeps it at a high pressure so the body stays firm.

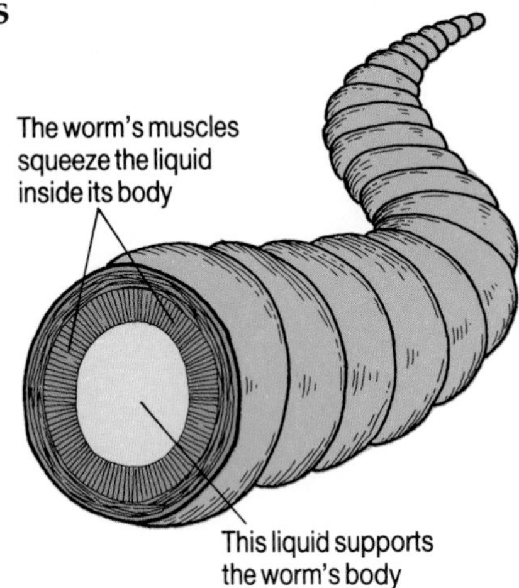

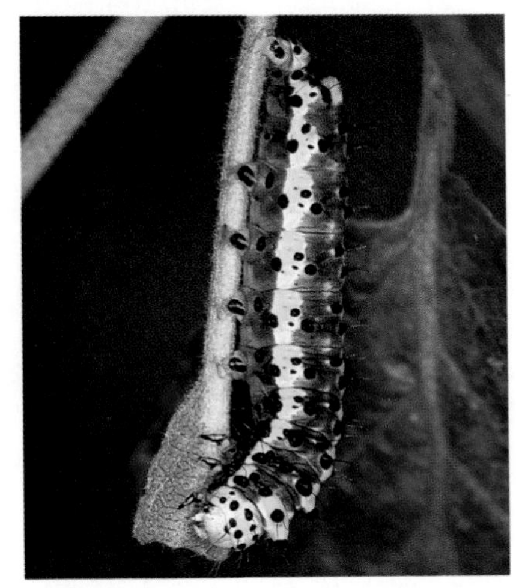
This will turn into a moth and get a hard external skeleton.

Animals with hard external skeletons

Insects, crabs, spiders, and other arthropods have a hard, tough skin over their bodies. This skin is like a suit of armour. It forms flat plates and hollow tubes which support and protect the body. This hard skin is called an **exoskeleton**.

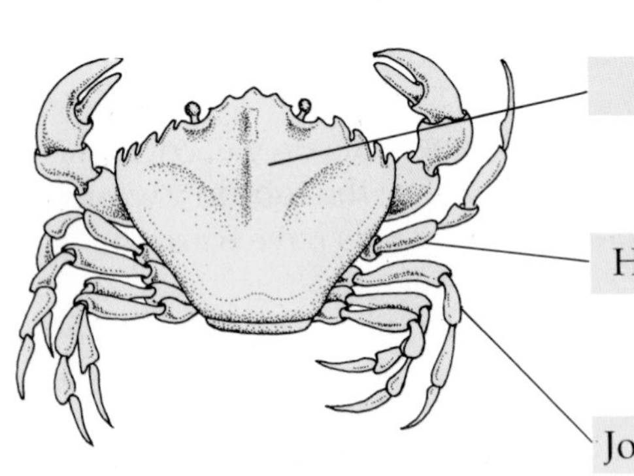

Flat plates cover the body.

Hollow tubes form the limbs.

Joints allow the limbs to move.

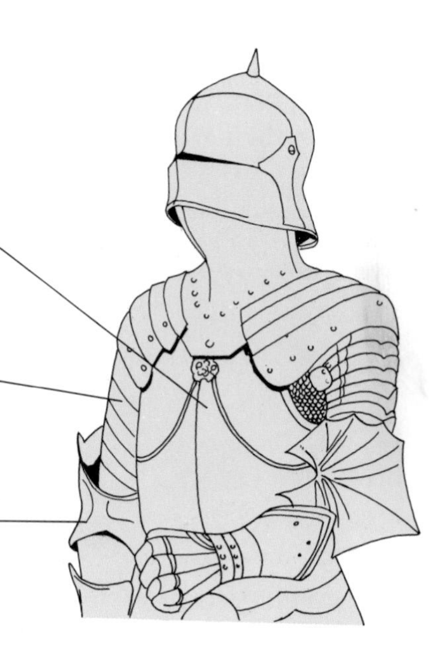

What an exoskeleton does

1 It **supports** the body and gives it shape.

2 It **protects** the soft insides from damage, dirt, and germs.

3 It is **waterproof** and stops the body drying up.

4 Its **colour** may help the animal to hide, or attract a mate.

5 Muscles are attached to it to move the joints.

A crab grows by shedding its exoskeleton. The skin beneath then hardens to form a new one.

A male stag beetle has antlers as part of its exoskeleton. It uses them to fight for females.

Animals with hard internal skeletons

You are supported by a hard skeleton inside your body, called an **endoskeleton**. It is made of bone. Fish, amphibia, reptiles, birds, and mammals all have endoskeletons.

The **skull** protects the brain, eyes, and inner ears.

Ribs protect the heart, lungs, and main blood vessels.

Muscles are attached to bones. They can pull on the bones and make them move.

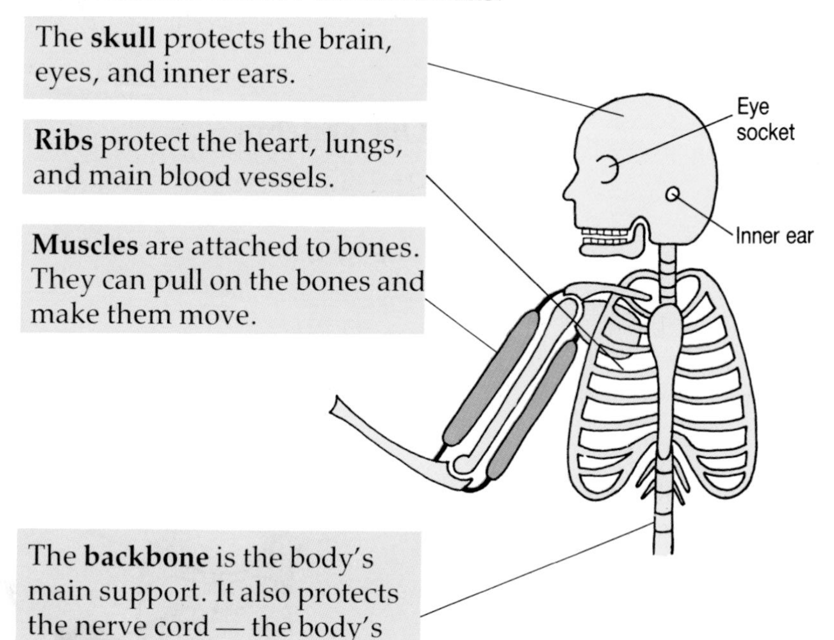

The **backbone** is the body's main support. It also protects the nerve cord — the body's main nerve.

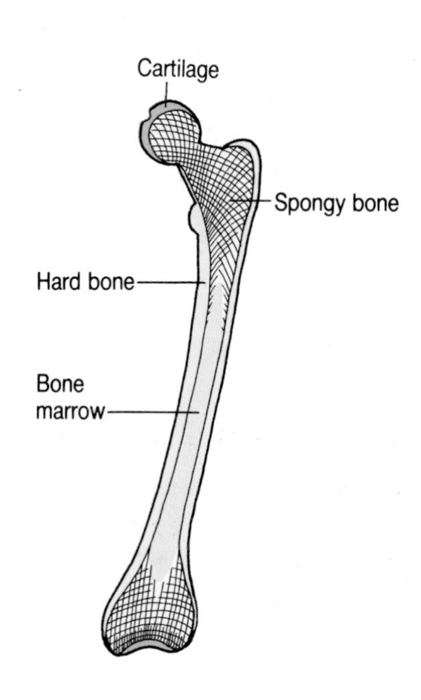

This is a leg bone cut in half. Bones are alive. They contain cells called **bone marrow** which make blood.

Questions

1 lobster mackerel housefly sparrow cat
Which of these have exoskeletons and which have endoskeletons?

2 What part of your skeleton protects:
a) your heart? **b)** your eyes? **c)** your nerve cord?

3 Does this describe an endoskeleton or an exoskeleton?
a) It stops the body drying up.
b) It makes blood cells.
c) It keeps dirt and germs out of the body.
d) It is waterproof.

4·2 More about the human skeleton

There are about 200 bones in your skeleton. This is a simple diagram showing the main ones.

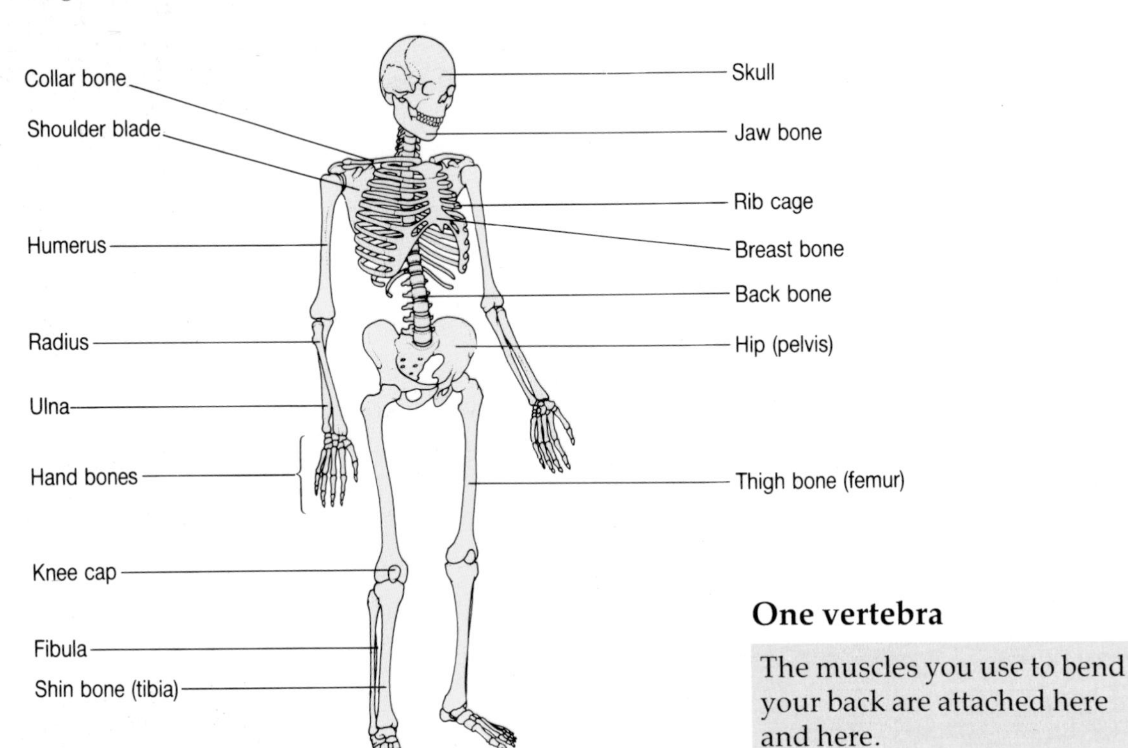

One vertebra

The muscles you use to bend your back are attached here and here.

This vertebra rubs against the next one here and here.

The vertebra has a hollow centre. The nerve cord goes through it.

The backbone

The backbone is called the **vertebral column**. It is made up of 33 small bones called **vertebrae**. These are joined in such a way that the backbone can bend and twist. This drawing shows part of a vertebral column seen from one side.

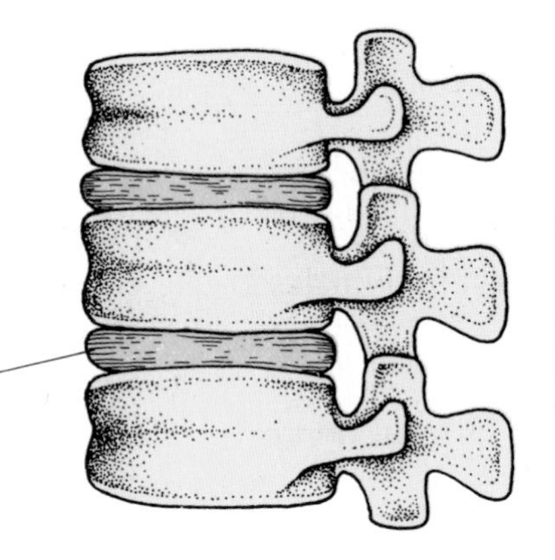

Intervertebral discs are pads of cartilage between the vertebrae. They stop vertebrae knocking against each other when you run or jump.

Joints

Joints occur wherever two or more bones touch. There are three kinds of joint:

1 fixed joints where the bones cannot move. There are fixed joints between the bones that make up the roof of your skull.

2 slightly moveable joints where the bones can move a little. There are slightly moveable joints between your vertebrae.

3 freely moveable joints where the bones can move easily. There are freely moveable joints at your knees and elbows.

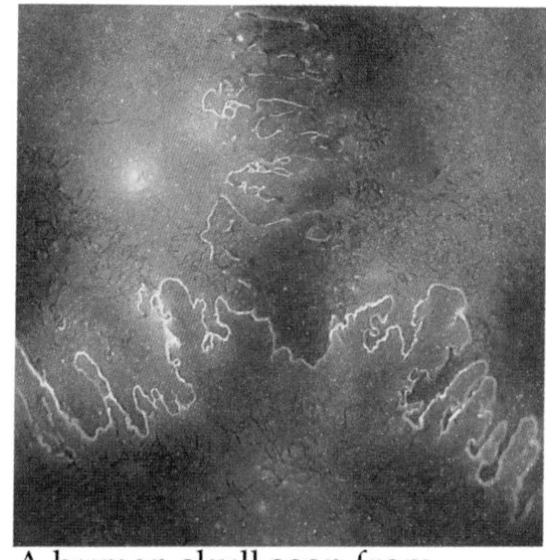

A human skull seen from above. The lines are the fixed joints.

Parts of a joint

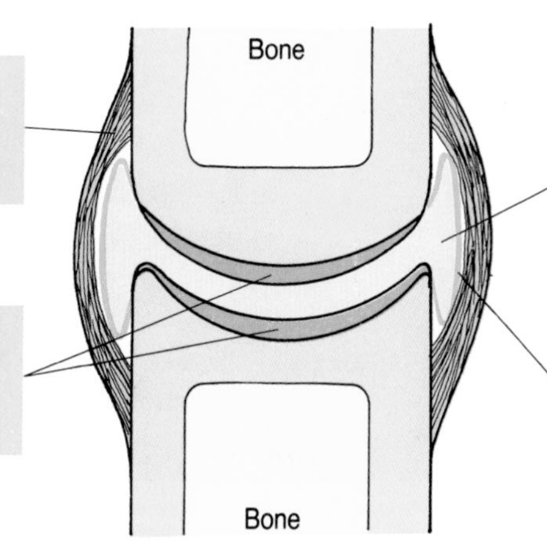

The bones at a joint are held together by strong fibres called **ligaments**.

Where the bones rub together they are covered by a slippery layer of **cartilage**.

An oily liquid called **synovial fluid** helps the joint move smoothly.

The **synovial membrane** around the joint stops the synovial fluid from draining away.

Two types of freely moveable joint

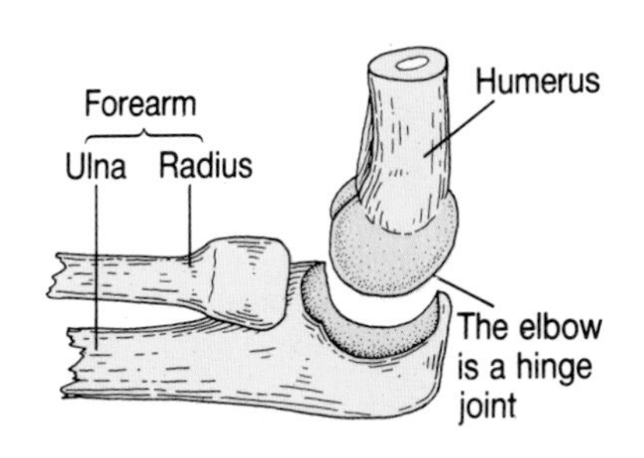

Hinge joints can bend in only one direction, like the hinge of a door. Your elbow is a hinge joint.

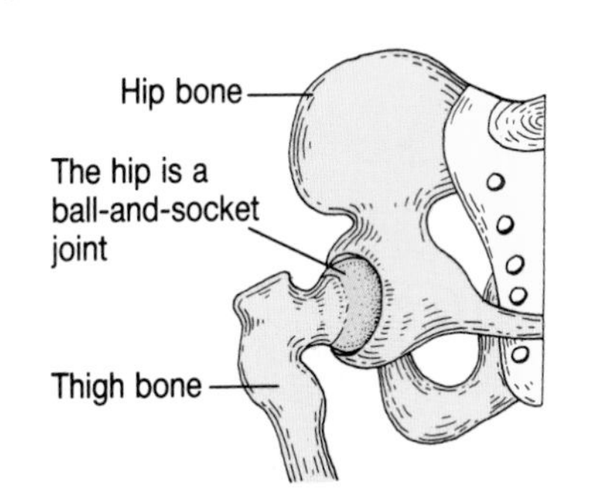

Ball-and-socket joints can bend in all directions. An example is the joint at your hip.

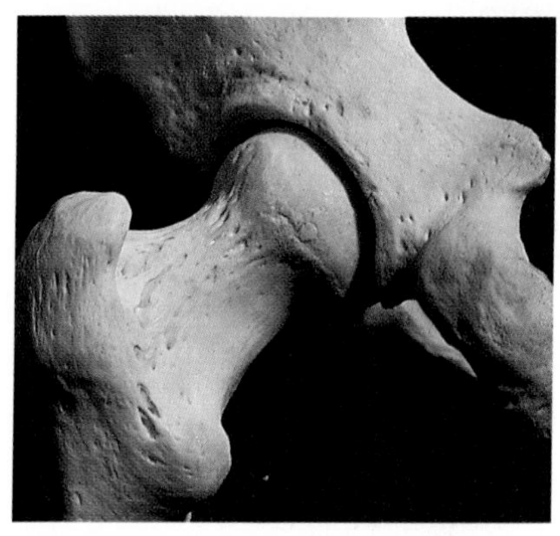

The hip joint from a human skeleton. Why is it called a ball and socket joint?

Questions

1. Write down the everyday names for these bones:
   a) pelvis **b)** vertebral column **c)** femur
   d) tibia
2. Where are your intervertebral discs, and what are they for?
3. What are ligaments for?
4. Name a hinge joint and a ball-and-socket joint on your body which are not mentioned here.
5. What is synovial fluid for?
6. What is cartilage for?

4·3 Muscles and movement

Muscles work by getting shorter, or **contracting**. When a muscle contracts it pulls what it is joined to, or squeezes something. There are three kinds of muscle in your body: **voluntary, involuntary** and **cardiac**.

Voluntary muscles

Voluntary muscles are muscles that you can control as you wish (voluntary means by free will).

The muscles attached to your bones are voluntary. When you decide to move they pull on your bones, which makes your joints bend. These muscles work quickly and powerfully, but they soon get tired.

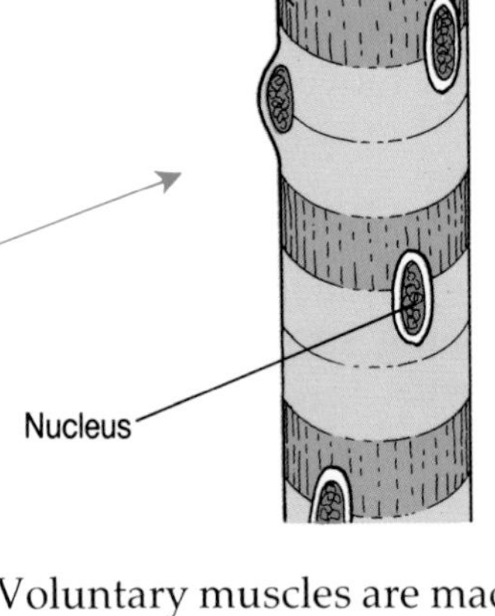

Voluntary muscles are made of fibres which are long and striped.

Involuntary muscles

Involuntary muscles are those you cannot control. They keep working on their own.

There are involuntary muscles in the walls of your gut. Their job is to push food along the gut. Involuntary muscles are also found in the walls of blood vessels, and the bladder. Involuntary muscles work slowly and do not get tired.

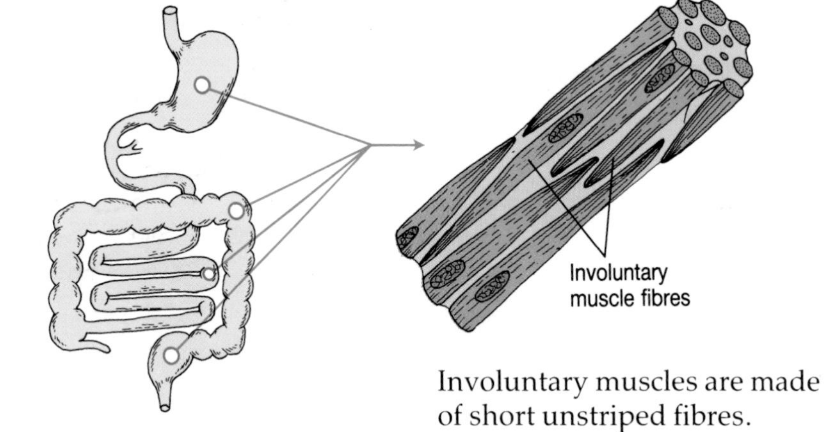

Involuntary muscles are made of short unstriped fibres.

Cardiac muscle

Your heart is made of cardiac muscle. This is a special kind of involuntary muscle.

Cardiac muscle works, without getting tired, all through your life, pumping blood around your body.

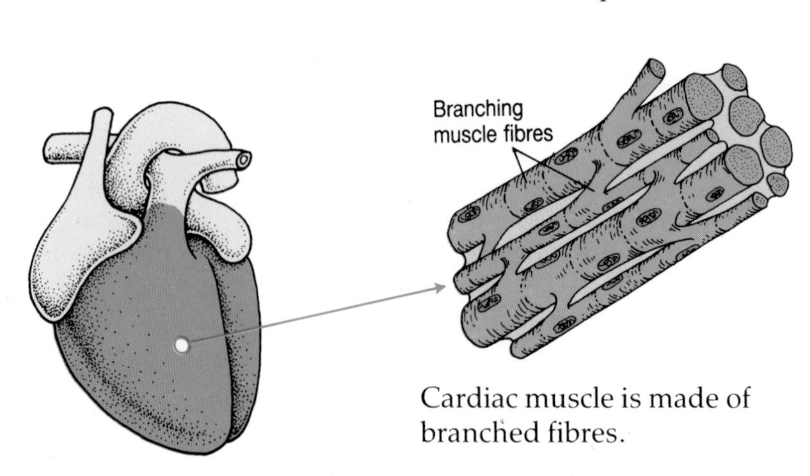

Cardiac muscle is made of branched fibres.

More about voluntary muscles

Muscles are attached to bones by strong fibres called **tendons**. A muscle has at least one tendon at each end.

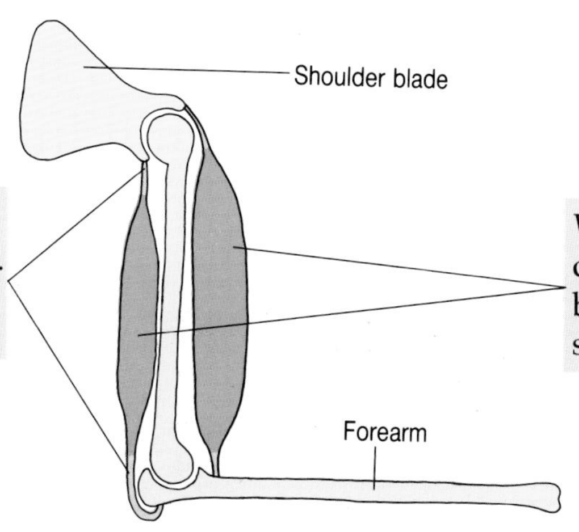

When one of these muscles contracts it pulls against the bones of the arm. This bends or straightens the elbow joint.

Voluntary muscles work in pairs

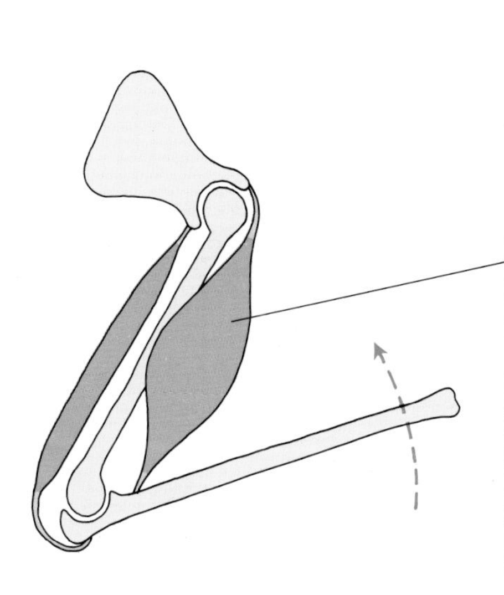

The muscles at a joint work in pairs. One muscle makes the joint bend, the other makes it straighten.

This muscle has contracted to bend or **flex** the elbow joint. Muscles which bend joints are called **flexor muscles**.

This muscle has contracted to straighten or **extend** the elbow joint. Muscles which straighten joints are called **extensor muscles**.

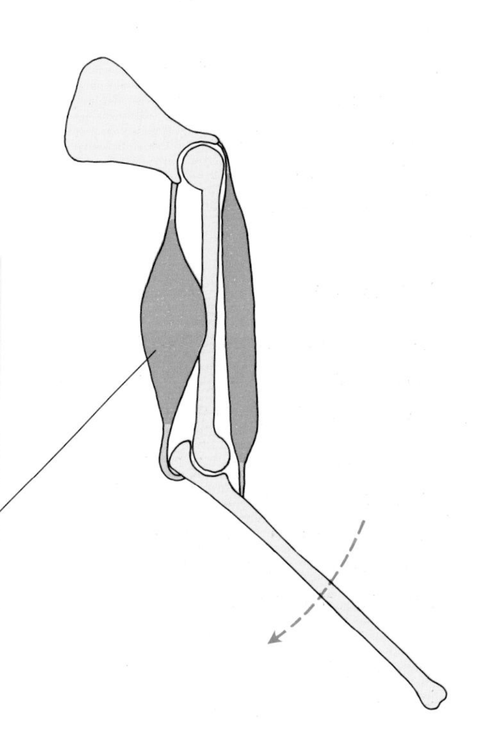

Questions

1 What kind of muscle:
   a) moves your arm when you turn this page?
   b) moves food from your mouth to your stomach?
   c) moves your legs when you run?
   d) moves blood around your body?

2 Why are at least two muscles needed at a joint?

3 What is the name for:
   a) muscles which straighten joints?
   b) muscles which bend joints?

4 Describe one way that heart muscle is:
   a) similar to stomach muscle.
   b) different from stomach muscle.

5 What is the difference between a ligament and a tendon?

4·4 Respiration

Even relaxing needs energy . . . and that means respiration. It goes on in all your cells even when you are quite still or asleep.

You use energy to walk, and think, and digest your food. In fact, you use energy for everything that goes on in your body. You get energy from food, during **respiration**.

Respiration goes on in the cells of every living thing.

Aerobic respiration

Respiration uses glucose from digested food. It usually uses oxygen too. Respiration which uses oxygen is called **aerobic respiration.** This is what happens:

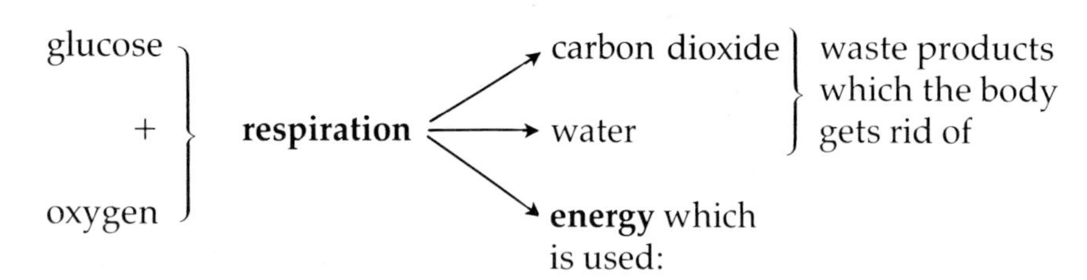

1. to work the muscles and other organs
2. to make the chemicals the body needs
3. to make new cells for growth, and to replace dead cells.

You can write aerobic respiration as:

glucose + oxygen → carbon dioxide + water + **energy**

Anaerobic respiration

Respiration can also take place without oxygen. It is then called **anaerobic respiration**. It produces less energy than aerobic respiration.

Microscopic organisms like yeast get energy by a type of anaerobic respiration called **fermentation**. It produces alcohol:

glucose → carbon dioxide + alcohol + **energy**

Alcoholic drinks are made using yeast and the sugar in fruit juice. Cider is made from yeast and apple juice. Beer is made from yeast and the malt obtained from barley seeds.

Fermentation is also used in baking. It takes place when yeast and sugar are added to dough. It produces carbon dioxide gas which fills the dough with bubbles and makes it rise. The yeast is killed in the hot oven.

Bubbles of carbon dioxide are produced when wine ferments.

Anaerobic respiration in muscles

Muscles can respire for a short time without oxygen (**anaerobically**). This happens when you run very fast. Your muscles need more energy than they can get from aerobic respiration. They get the extra energy they need from **anaerobic respiration**. This produces **lactic acid** instead of carbon dioxide.

glucose → lactic acid + **energy**

Lactic acid gathers in your muscles making them ache so much that you soon have to stop running. You get rid of the lactic acid by rapid breathing after you stop. This supplies your muscles with oxygen, which combines with lactic acid to make carbon dioxide and water.

The amount of oxygen needed to get rid of lactic acid from muscles is called the **oxygen debt**.

After an exhausting race, these rowers gasp for breath to pay back the oxygen debt.

The respiratory system in humans

You use your **respiratory system** to breathe in oxygen for respiration, and to breathe out carbon dioxide produced by respiration.

The voice box or **larynx**. It makes sounds used in speaking.

The wind-pipe or **trachea**. It is like the flexible hose of a vacuum cleaner or tumble drier. It is held open by rings of **cartilage**.

The **lungs** are soft and spongy.

The **ribs** protect the lungs.

The **intercostal muscles** help with breathing in and out.

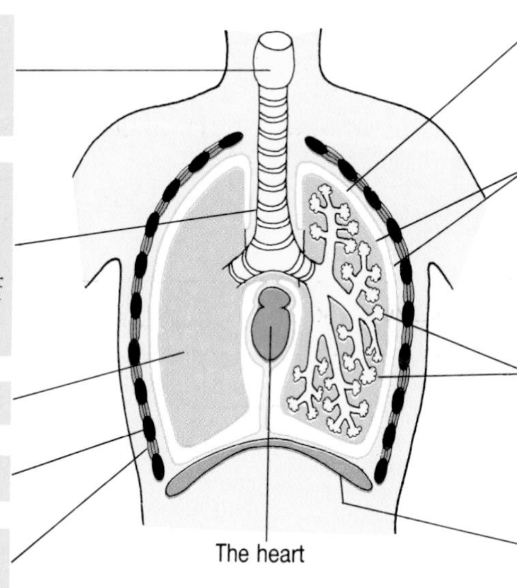

Lungs are in a space in the chest called the **thoracic cavity**.

This cavity is lined with a slippery skin called the **pleural membrane**. It protects the lungs as they rub against the ribs.

The wind pipe has thousands of branches which end in tiny **air sacs**. This is where oxygen is taken in to the body, and carbon dioxide is got rid of.

The **diaphragm** is a sheet of muscle below the lungs. It helps with breathing in and out.

Questions

1 Why is respiration essential for life?

2 What is the difference between aerobic and anaerobic respiration?

3 What waste substances are produced by aerobic respiration?

4 Which kind of respiration is going on in your body now, and which when you run very fast?

5 How does your body use the energy from food?

6 Why has the wind-pipe got rings of cartilage?

7 What is the larynx for?

4·5 Lungs and gas exchange

Your body takes in oxygen from the air for respiration. At the same time it gives out the carbon dioxide produced by respiration. This exchange of gases takes place in your lungs.

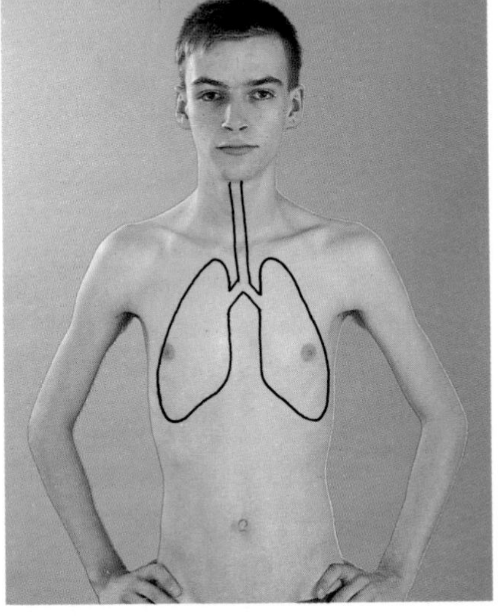

Where your lungs are . . .

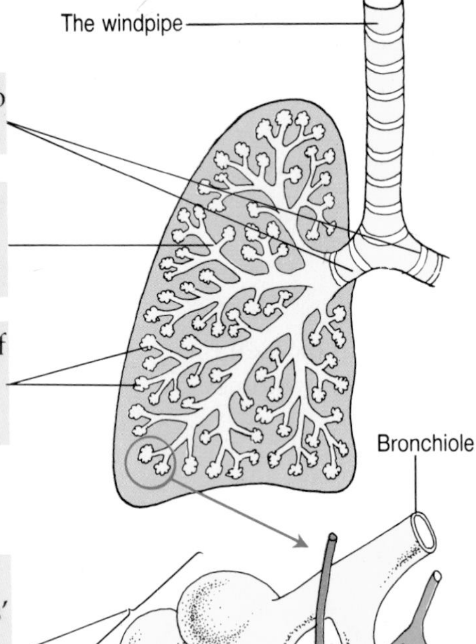

The wind-pipe divides into two tubes called **bronchi**.

The bronchi divide thousands of times into narrower tubes called **bronchioles**.

There are **air sacs** at the ends of the bronchioles. This is where gas exchange takes place.

A closer look at air sacs

This is one **air sac**. It looks like a bunch of grapes. The 'grapes' are called **alveoli**. They are smaller than grains of salt, and there are 300 million of them in your lungs.

Alveoli have thin moist walls, so that gases can pass through them easily.

Alveoli are covered with narrow blood vessels called **capillaries**. Oxygen passes into the capillaries from the alveoli.

Gas exchange in the alveoli

This diagram shows what happens in the alveoli.

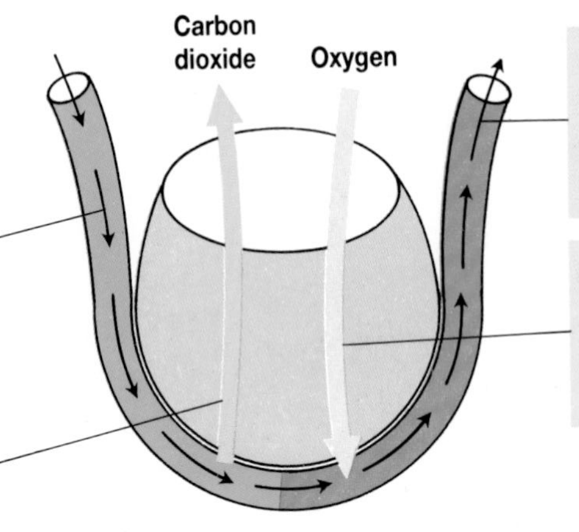

Blood flows to the lungs from around the body. It carries carbon dioxide produced by respiration in the cells of the body.

Carbon dioxide passes *from* the blood *into* the alveoli. Then it is breathed out of the body.

Blood carries oxygen away from the lungs to every cell in the body, where it is used for respiration.

Oxygen is breathed into the lungs. It dissolves in the water lining the alveoli. From there it passes *into* the blood.

When you breathe in

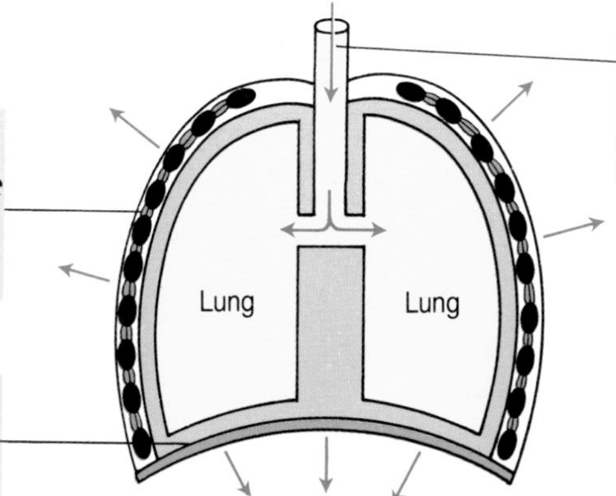

1 The **intercostal muscles** contract. These pull the rib cage upwards. So the chest increases in volume.

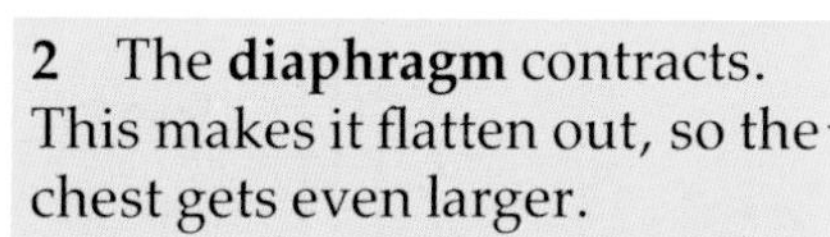

2 The **diaphragm** contracts. This makes it flatten out, so the chest gets even larger.

3 As the chest gets larger, air is sucked down the wind-pipe and into the lungs.

When you breathe out

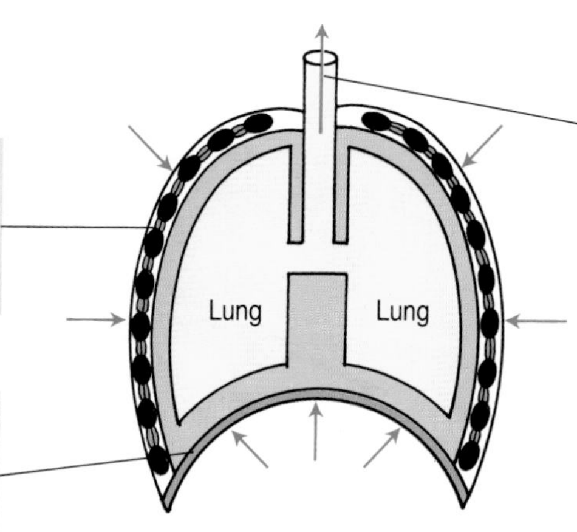

1 The **intercostal muscles** relax, which lowers the rib cage. The chest decreases in volume.

2 The **diaphragm** relaxes, and bulges upwards. This decreases the volume of the chest even more.

3 Because the chest has got smaller, air is forced out of the lungs.

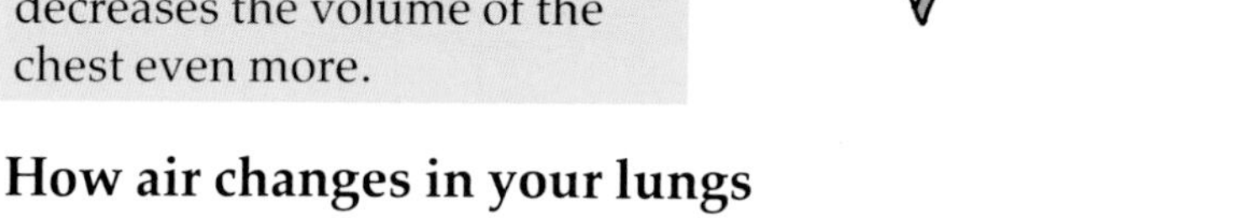

How air changes in your lungs

Gas	Amount of it in air you breathe in	Amount of it in air you breathe out
Oxygen	21%	16%
Carbon dioxide	0.04%	4%
Nitrogen	79%	79%
Water vapour	a little	a lot

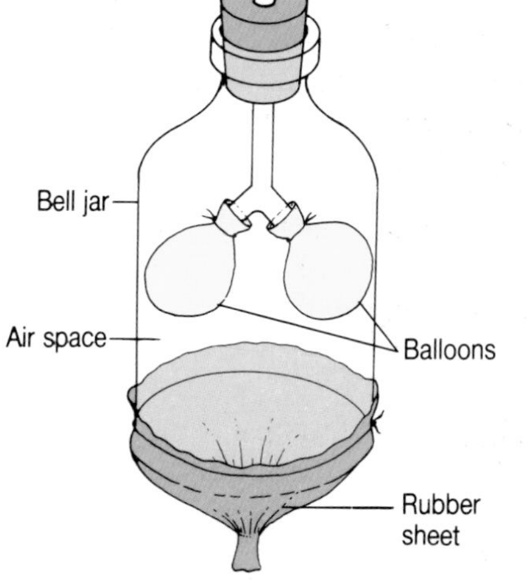

Watch what happens in a **bell jar** when you pull the rubber sheet downwards.

Questions

1. Which gases are exchanged in your lungs?
2. Where in the lungs does gas exchange take place?
3. Explain what happens to intercostal muscles and the diaphragm as you breathe in and out.
4. Explain why you breathe out:
   a) more carbon dioxide than you breathe in
   b) as much nitrogen as you breathe in
5. Do you use *all* the oxygen your breathe in?
6. What are alveoli?

Questions

1 This diagram shows muscles which control the knee joint.

a) Name the parts labelled A to E (page 69 should help.)

b) What is the main difference between voluntary and involuntary muscles? Give one example of each.

c) Are muscles X and Y voluntary or involuntary muscles?

d) What do flexor muscles and extensor muscles do?

e) Look at muscles X and Y in the diagram. Which is an extensor muscle, and which is a flexor muscle?

f) Name a hinge joint and a ball-and-socket joint on this diagram.

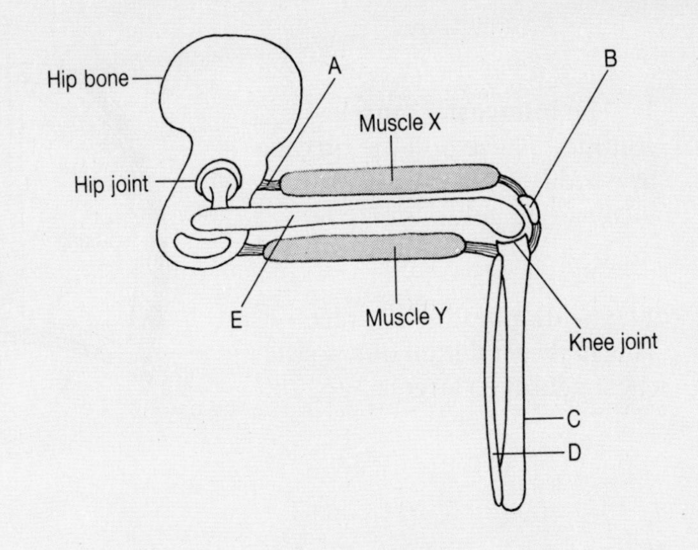

2 This diagram shows a freely moveable joint. Match each label on the diagram with one or more of the descriptions below:

a) an oily liquid;

b) slippery gristle;

c) fibres which hold the bones together;

d) synovial membrane;

e) lubricates the joint;

f) ligament.

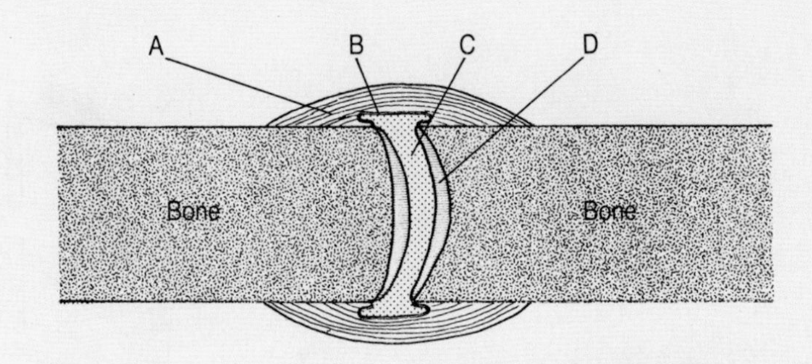

3 This diagram shows the parts of the human respiratory system.

Match each label on the diagram with one or more of the descriptions below:

a) holds open the wind-pipe;

b) the trachea;

c) bulges upwards when it relaxes;

d) ribs;

e) bronchus;

f) contain millions of alveoli;

g) tube connecting the mouth with the lungs;

h) lungs;

i) diaphragm;

j) lift the rib cage when they contract.

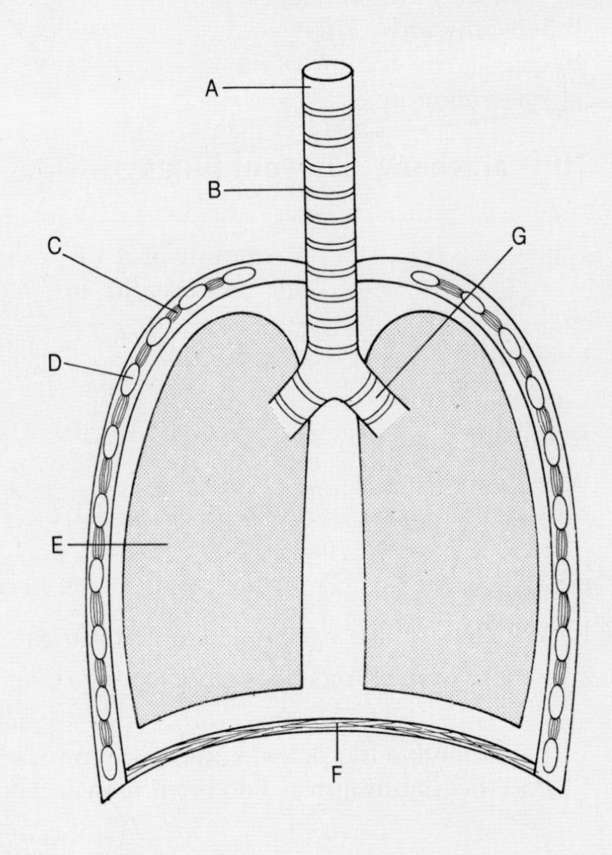

4 a) Name the gas your muscles produce when they are working.

b) How does your body get rid of this gas?

c) Name the gas your muscles need more of, during exercise.

d) How does this gas get to your muscles?

1 Exoskeletons

a) Catch a beetle. (They live under stones and pieces of rotting wood, and in compost heaps). Obtain a crab from a fish shop.
b) Compare the exoskeleton of the crab with that of the beetle. In what ways are they alike? In what ways are they different?
c) Look at the joints of the crab's leg. Notice how each joint moves in a slightly different direction.
d) Watch the beetle moving to see how legs like this work. Then let the beetle go where you found it.
e) Write down at least two differences between a crab's legs and your own legs.

2 How many bones have you got?

a) Copy the chart below. Fill in the missing numbers by feeling your bones through your skin. (Page 66 should also be useful.) Then add up the numbers to find the total.

Parts of the skeleton	Number of bones
Skull	21
Lower jaw	
Backbone	33
Shoulder blades	
Collar bones	
Breast bone	7
Ribs	
Hips	6
Arms	
Wrists	16
Hands	
Legs	
Knee caps	
Ankles	14
Feet	
TOTAL	

3 Muscles

Roll up one sleeve so that you can see the muscles of your upper arm. The muscle at the front of your arm above the elbow is the **biceps muscle.** The muscle at the back of your arm is the **triceps muscle.**

a) With your arm hanging by your side, feel both muscles. They are both soft.
b) Press down on the desk top with your hand. Which muscle feels hard, the biceps or triceps?
c) Hook your fingers under the edge of the desk and pull gently upwards. Which muscle is hard now, your biceps or triceps?
d) Which of these two muscles is the flexor muscle of your arm, and which is the extensor muscle?

4 Respiration

Set up the apparatus shown in the diagram below.

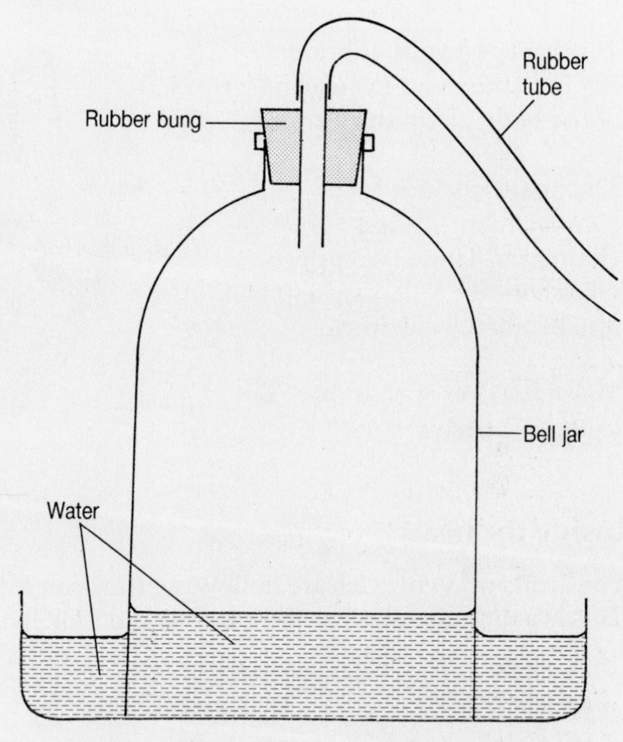

a) Take out the rubber bung and pass a lighted taper through the mouth of the bell jar. Notice how long it takes for the flame to go out.
b) Lift the bell jar out of the water to clear smoke and fumes. Put back the bung and rubber tube.
c) Put the end of the tube in your mouth, and breathe the air in the bell jar through it several times.
d) Remove the bung and tube and immediately push a lighted taper into the bell jar.
e) How long does the flame last this time?
f) What gas did your body remove from the air in the bell jar, while you were breathing it?
g) How does this investigation show that the gas was removed?

5·1 The heart

Clench your fist and look at its size. Your heart is about the same size. It is made of special muscle called **cardiac muscle**. Its job is to pump blood around your body. It pumps about 40 million times a year and weighs about as much as a grapefruit.

The outside of a heart looks like this.

This tube is a **vein**. It brings blood to the heart from all parts of the body *except* the lungs.

The heart has four compartments called **chambers**. These two upper chambers are called **atria**. (Each one is called an **atrium**.)

These two lower chambers are called **ventricles**.

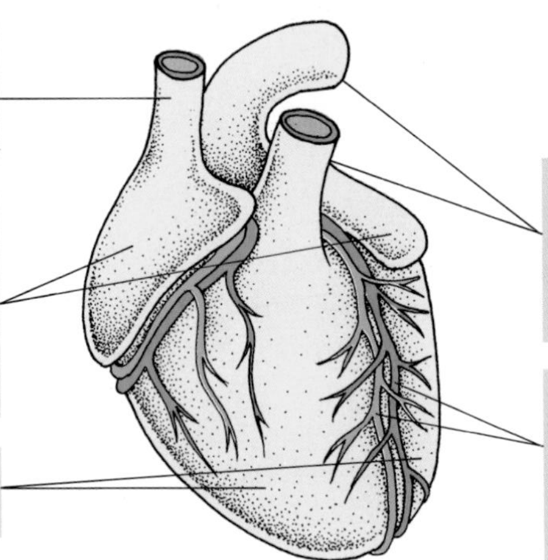

These two tubes are called **arteries**. They carry blood away from the heart to all parts of the body.

The heart has its own blood supply, carried by the **coronary** artery and vein. These tubes carry food and oxygen to the heart, and carry wastes away.

Inside the heart

The atria and ventricles are hollow, so they can fill up with blood. This is a diagram of what the heart would look like sliced open.

This **artery** carries blood to the lungs.

This **valve** stops blood flowing back into the heart.

This **vein** carries blood into the heart from the body.

The **right atrium** has thin walls.

This **valve** allows blood to flow from the right atrium to the right ventricle only.

These **valve tendons** are strings which hold valve flaps in place.

The **right ventricle** has thick walls. It pumps blood to the lungs.

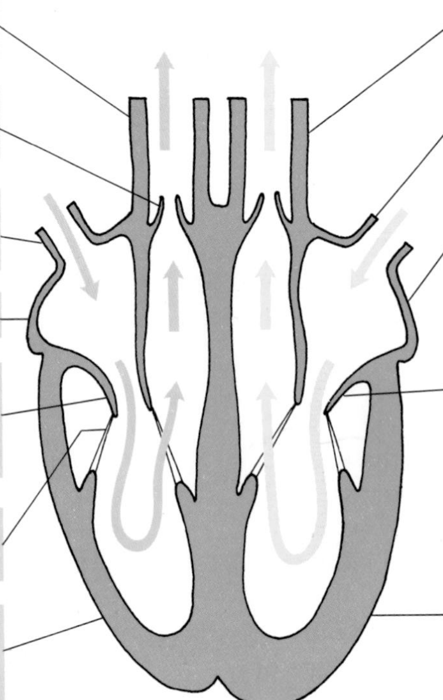

This **artery** carries blood away from the heart to the body.

This **vein** carries blood into the heart from the lungs.

The **left atrium** has thin walls.

This **valve** allows blood to flow from the left atrium to the left ventricle only.

The **left ventricle** has very thick walls. It pumps blood to all parts of the body, except the lungs.

How the heart pumps blood

The heart pumps blood by tightening, or **contracting.** That makes it smaller, so blood gets squeezed out into the arteries. Then it **relaxes** again, and fills up with blood from the veins.

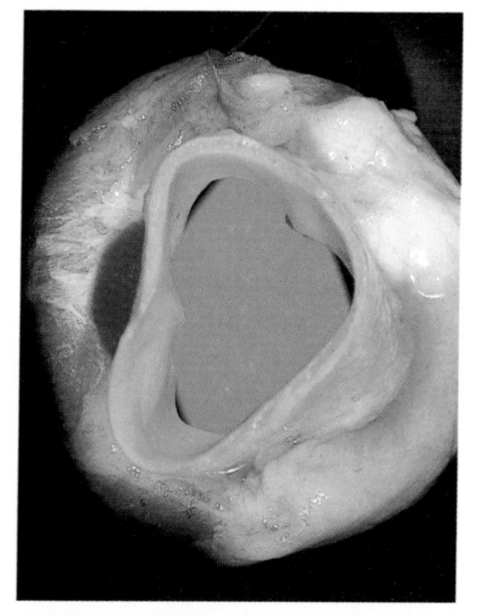

This photograph shows the valve in an artery near the heart. The valve is open, so blood can flow through.

1 When the heart is relaxed, both sides fill up with blood from the veins.

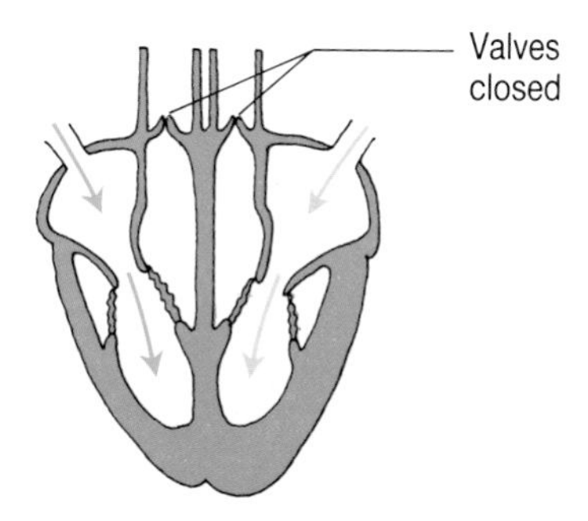

2 Then the **atria** contract. The veins also contract where they join the atria. So blood is forced into the ventricles through the valves.

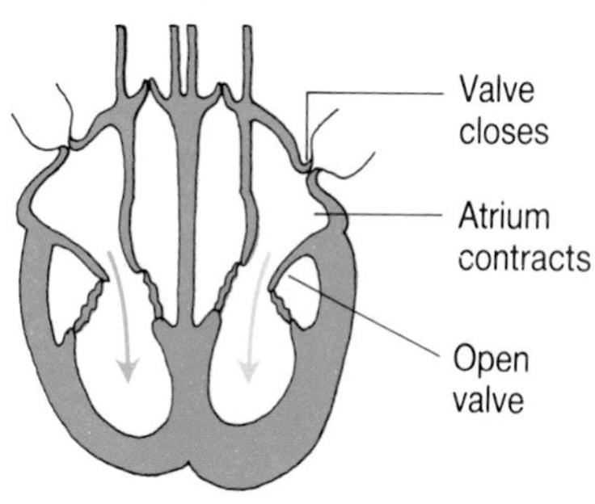

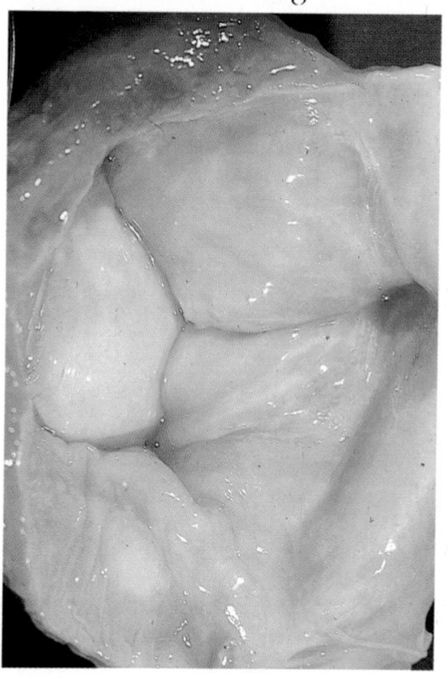

This shows the same valve closed to stop blood flowing backwards into the heart.

3 A fraction of a second later, the **ventricles** contract. The valves between the atria and ventricles close. So the blood is squeezed into the arteries.

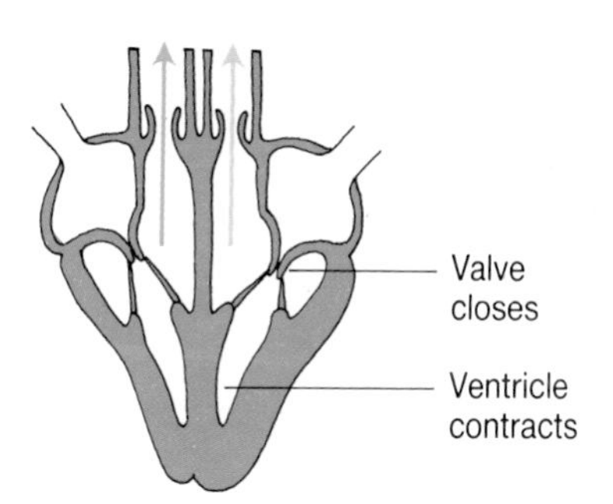

4 The heart relaxes again, and fills up with blood.

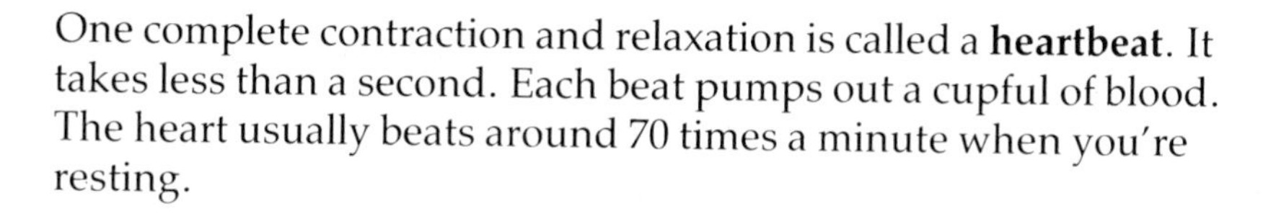

One complete contraction and relaxation is called a **heartbeat**. It takes less than a second. Each beat pumps out a cupful of blood. The heart usually beats around 70 times a minute when you're resting.

Questions

1 What is the heart made of?
2 What are the upper parts of the heart called?
3 Which ventricle has the thickest walls?
4 What stops blood flowing backwards through the heart?
5 What are valve tendons for?
6 What are the coronary arteries and veins for?
7 The heart can **contract.** What does this mean?
8 What happens when the atria contract?
9 What happens when the ventricles contract?
10 What is a heartbeat?
11 How fast does the heart usually beat?

5·2 What makes blood flow

When the powerful muscles of your heart contract they force blood out into tubes called **arteries**. The arteries branch into tiny little tubes called **capillaries**. The capillaries join together to form **veins**. The veins carry blood back to the heart.

This is an **artery**. There are elastic fibres in its walls. The blood pumps into the artery very fast, at high pressure, so the elastic fibres stretch. Then they contract, and that squeezes the blood towards the capillaries.

This is a **capillary**. Its walls are so thin that liquid from the blood can pass through them. This liquid takes food and oxygen to the cells of the body. It also takes away carbon dioxide and other wastes.

This is a **vein**. Veins are wider than arteries and have thinner walls. The blood flows through them more slowly. They have valves to stop it flowing backwards.

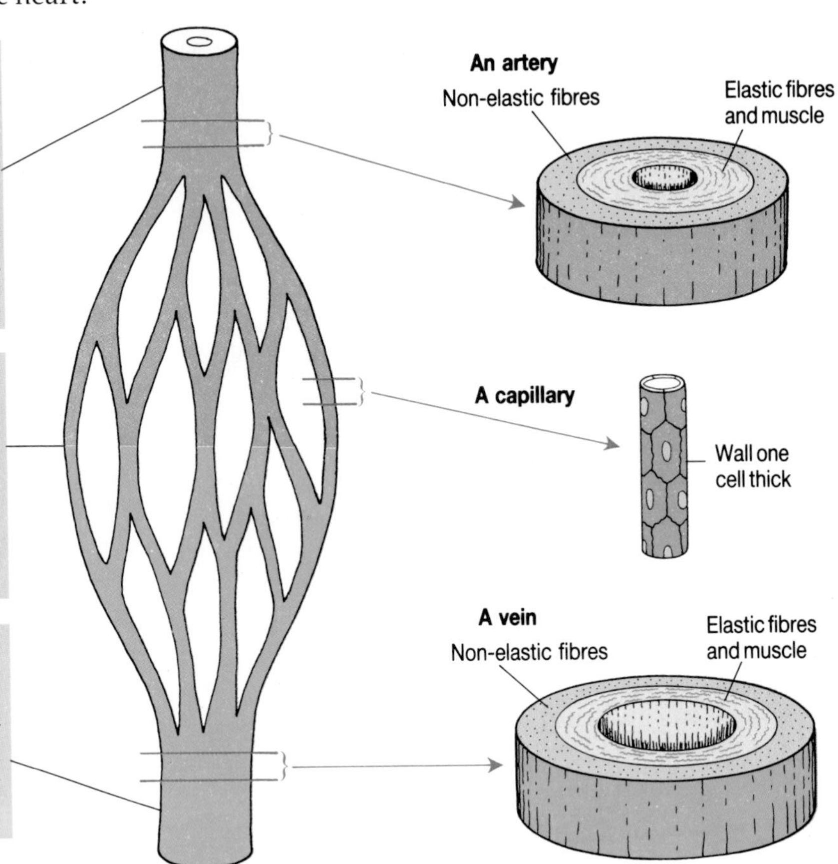

Vein valves

There are many large veins inside muscles of the legs and arms. When these muscles contract they squeeze the veins. This squirts blood towards the heart. The valves stop it flowing in the opposite direction.

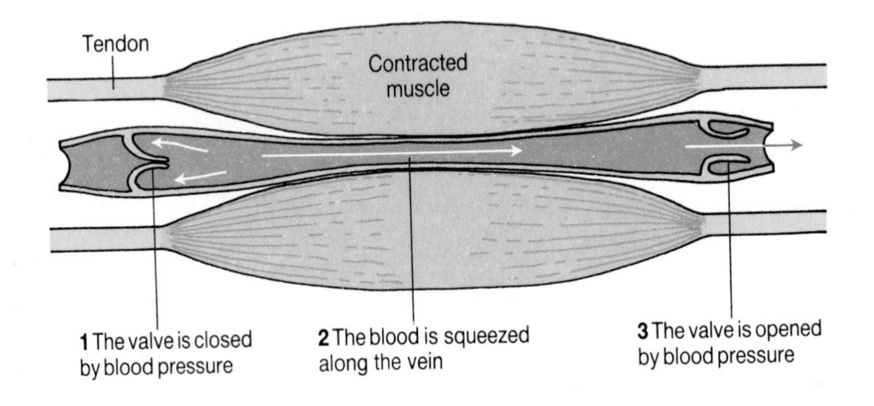

The circulatory system

The tubes that carry blood are called **blood vessels**. The heart and blood vessels together make up the **circulatory system**. This system has two parts:

1 The right side of the heart pumps blood to the lungs and back again. In the lungs it loses carbon dioxide and picks up oxygen.

2 The left side of the heart pumps blood to the rest of the body and back. On its way around the body the blood loses oxygen to the body cells and picks up carbon dioxide.

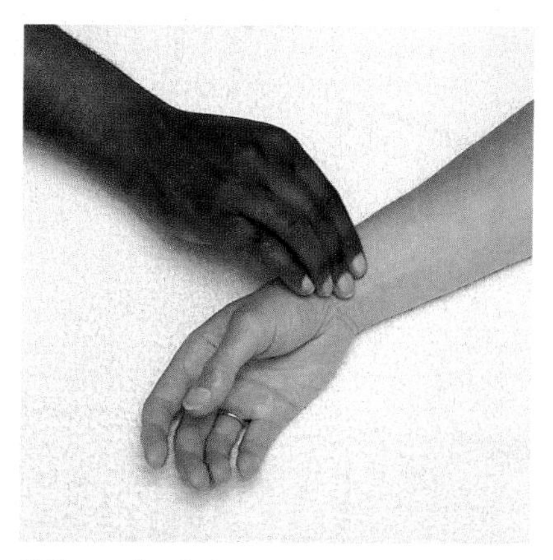

When the heart pumps blood into the arteries at high pressure, it causes a pulse. Use your fingers, as shown here, to find the pulse at the wrist.

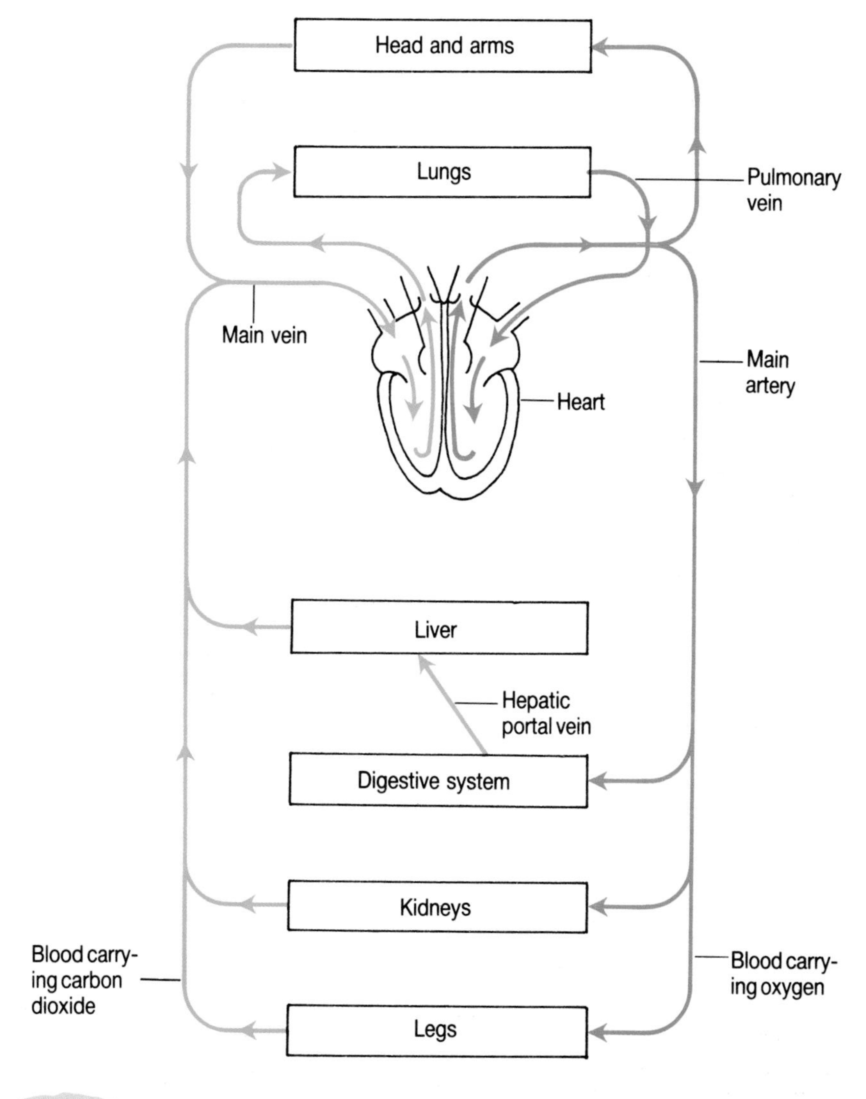

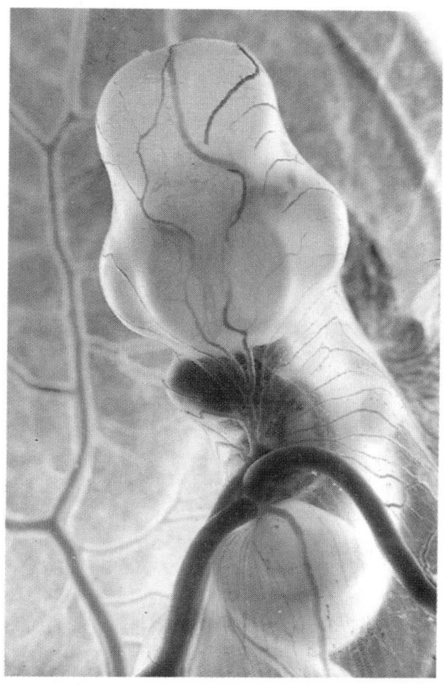

This photograph shows the network of capillaries and arteries in a chick embryo. There are enough capillaries in your body to stretch one-and-a-half times round the world!

Questions

1 What happens in the arteries?

2 What happens in the capillaries?

3 How are arteries different from veins?

4 **a)** Blood is at its highest pressure just where it leaves the heart. Why do you think this is?

b) The blood in an artery flows faster and at higher pressure than the blood in a vein. Can you explain why?

5 In the diagram above find:

a) a vein with blood full of oxygen

b) an artery containing blood full of oxygen

c) a vein with blood full of digested food.

5·3 Heart disease

To keep pumping, your heart needs food and oxygen. It gets them from its own blood supply, carried in the **coronary arteries**. If these arteries get blocked the result is **heart disease**.

How arteries get blocked

A fatty substance called **cholesterol** can stick to the walls of an artery. The arteries become narrower. So blood gets slowed down.

Cholesterol can make artery walls rough. This causes blood to clot, as it flows past. A blood clot can block an artery completely. The blockage is called a **thrombosis**. The blood flow is stopped.

Bits of cholesterol can break off into the blood stream, and block narrow blood vessels.

A thrombosis in blood vessels in the brain is called a **stroke**. Brain cells die. A person suffering from a stroke may get paralysed or even die.

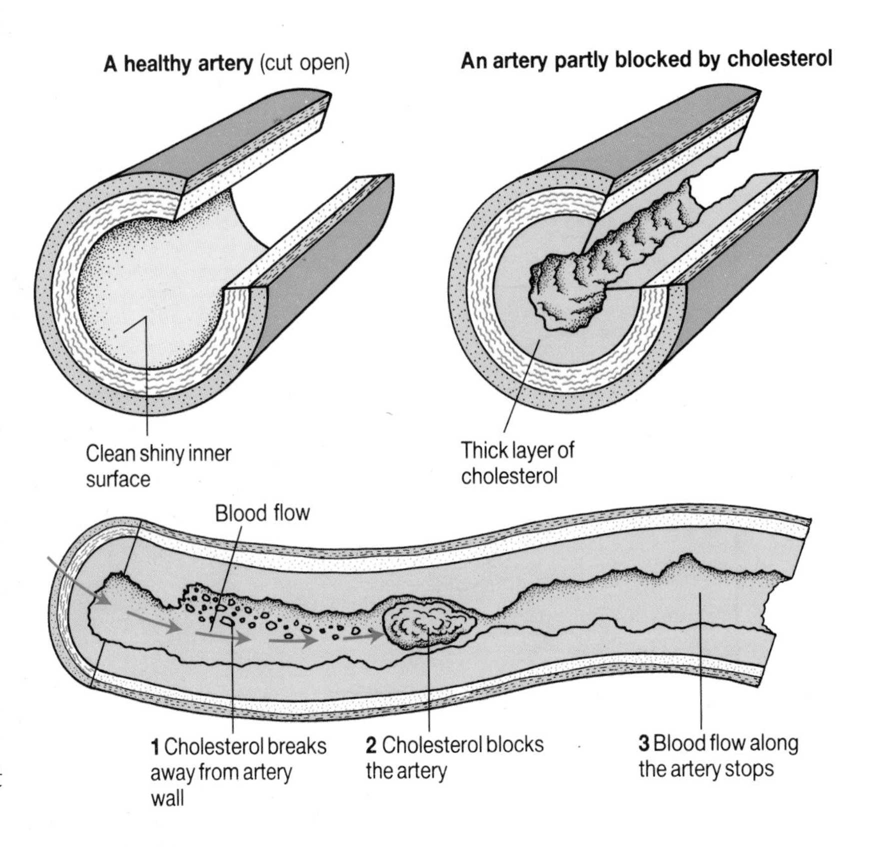

Blocked coronary arteries

If a coronary artery gets partly blocked, the heart muscle gets too little food and oxygen. The result is severe chest pains, called **angina**.

A thrombosis in the coronary artery is called a **heart attack**. The heart stops beating. But with the right treatment, the heart can be forced to start beating again.

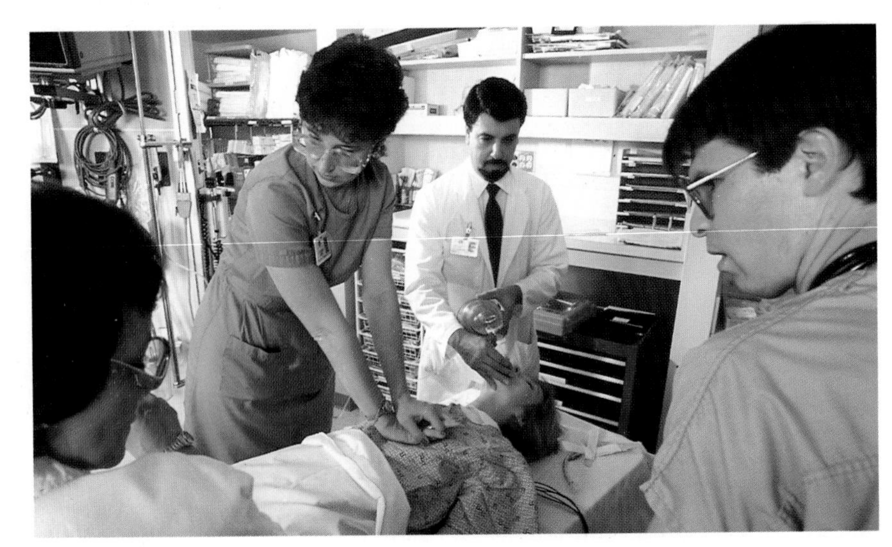

Emergency treatment for a heart attack. The victim's heart is massaged to start it beating again. The green cylinder contains oxygen.

Things that can lead to heart disease

Heart disease is a big problem. In Britain it kills about 400 people a day. Most doctors agree that your chances of suffering from it are greatest:

1 if you eat lots of cream, butter, eggs, fat, and fried foods. These can lead to a high level of cholesterol in the blood
2 if you smoke cigarettes
3 if you are overweight
4 if you take little or no regular exercise
5 if your way of life involves stress (worry, anger, fear and so on).

One way to avoid heart disease is to take regular exercise.

How you can avoid heart disease

1 Cut down on fried food. You can grill, boil, or steam, rather than fry. If you do fry food use corn, soya, or sunflower oils.

2 Eat less red meat. When you do eat it cut off any fat you can see.

3 Eat less dairy foods (eggs, butter, milk, and cream).

4 Eat more poultry and fish, because these are less fatty.

5 Eat more fresh fruit and vegetables.

6 Do not smoke.

7 Take exercise regularly.

8 Take time to relax before you go to bed.

If you follow these guidelines you will be less likely to suffer from heart disease. Your general health will be better too.

Questions

1 Name the fatty substance that can block arteries.

2 What is a thrombosis? What causes it?

3 What is a stroke?

4 What causes the disease called angina?

5 Explain what causes a heart attack.

6 **a)** Why is the man in the drawing above likely to develop heart disease?
b) Name two things, not shown in this drawing, which can also cause heart disease.

7 How can you avoid heart disease?

5·4 What blood is

You have nearly half a bucket of blood (5.5 litres) in your body! Blood is a liquid called plasma, with red cells, white cells, and platelets floating in it. Let's look at each of these things in turn.

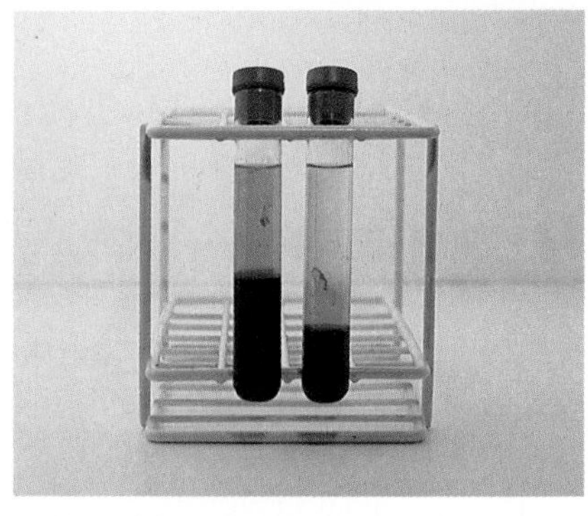

When blood is left standing for a time, the solid particles start settling to the bottom of the plasma.

Plasma

Plasma is a yellow liquid. It is mainly water, with digested food, hormones, and waste substances dissolved in it.

Red cells

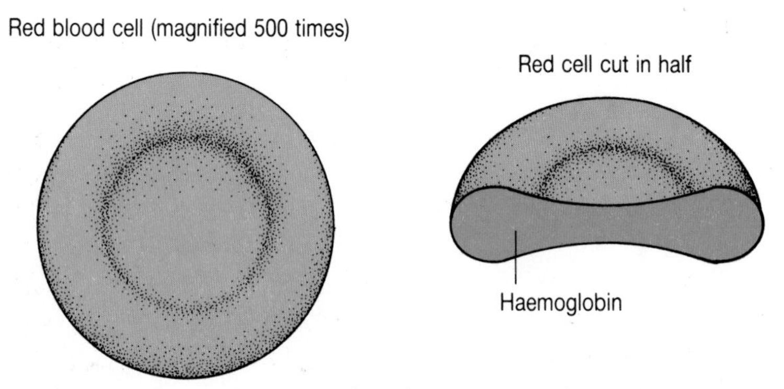

Red cells are wheel-shaped. They have no nucleus. They are red in colour because of the **haemoglobin** inside them.

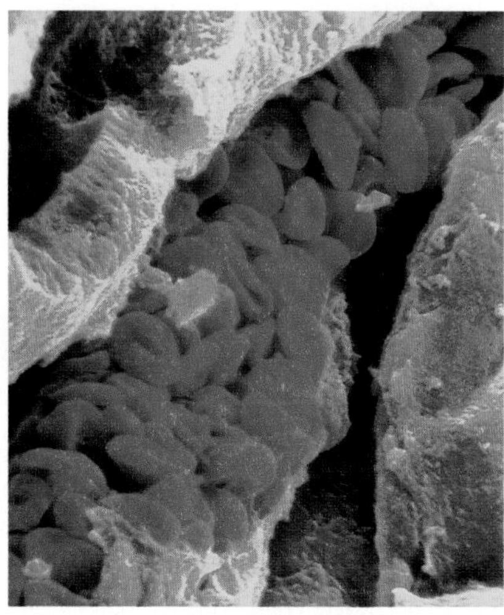

There are about five million red cells in one drop of blood. This photograph shows them moving through a capillary.

How the red cells carry oxygen

Red cells are the body's oxygen carriers. They carry oxygen from the lungs to all the cells of the body.

1 The red cells pick up oxygen as blood is pumped through the lungs.

2 The oxygen and haemoglobin join to form **oxyhaemoglobin**. This is bright red.

3 As the blood passes around the body, the oxyhaemoglobin breaks down and releases oxygen to the body cells.

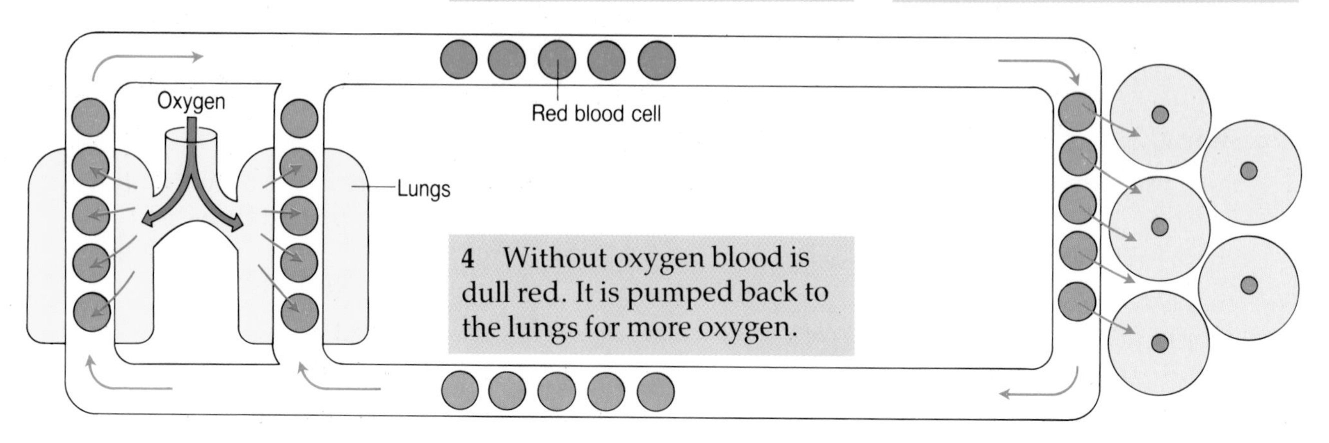

4 Without oxygen blood is dull red. It is pumped back to the lungs for more oxygen.

White cells

White cells are larger than red cells. They all have a **nucleus.** They can change shape. White cells protect us from disease. White cells called **phagocytes** can eat up the germs that cause disease.

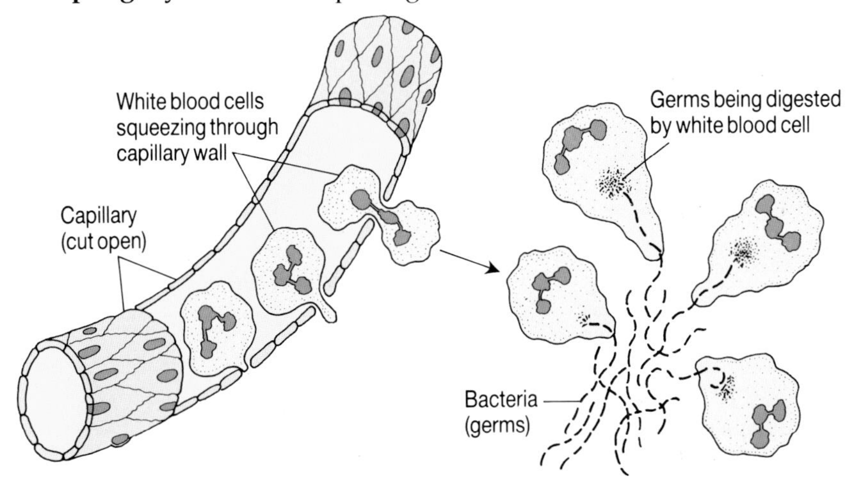

1 Phagocytes can squeeze through capillary walls.

2 They move towards germs and surround them. They then digest them.

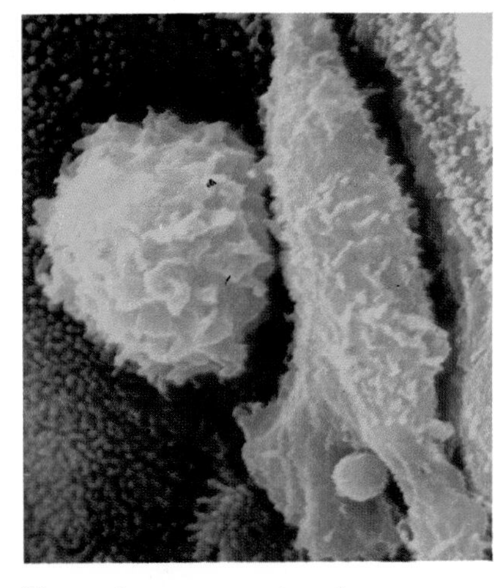

Two phagocytes in a human lung. One has grown long and thin, and is about to destroy a small particle. Phagocytes keep lungs clear of dust and pollen.

Other white cells make chemicals called **antibodies.** These chemicals destroy germs that get into the body by making them stick together, or by dissolving them. They also destroy **toxins** (poisons) that germs make. There is a different antibody for each kind of germ.

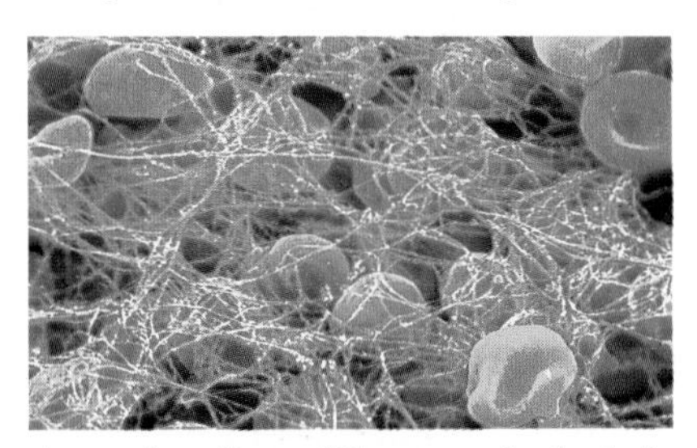

A cut healing. The tangled pink fibres were produced by the platelets. The orange bits are trapped red cells.

Platelets

Platelets are fragments of cells formed in bone marrow. Their job is to help stop bleeding from cuts.

1 Bleeding washes out dirt and germs from the cut. Then the platelets produce tiny fibres. Red cells get trapped in these fibres and the blood changes into a thick red jelly called a blood clot.

2 The clot hardens to a scab. This keeps the wound clean while new skin grows. Then the scab breaks off.

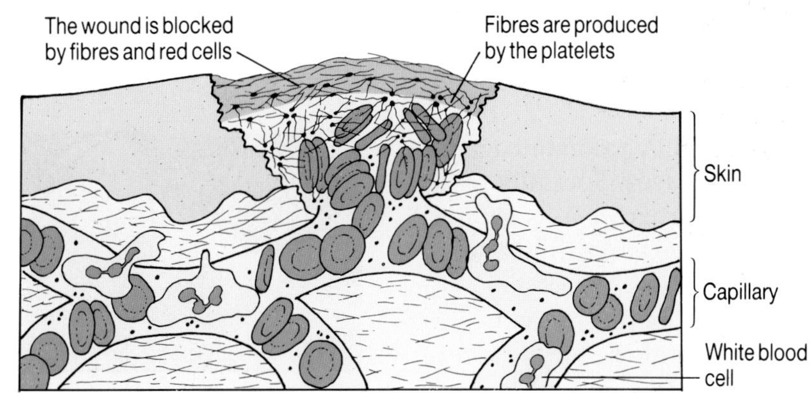

Questions

1 Name the four things that make up blood.

2 Name the liquid part of blood.

3 What do red cells do?

4 What are phagocytes?

5 What are antibodies?

6 How do platelets help blood clot?

5·5 What blood does

Blood does two jobs. It carries things round the body, and it protects us against disease.

Blood carries things

What it carries	How carried
1 Oxygen from the lungs to the rest of the body	In the red cells
2 Carbon dioxide from the body to the lungs	Mainly in plasma
3 Digested food from the gut to the rest of the body	In the plasma
4 Wastes from the liver to the kidneys	In the plasma
5 Hormones from hormone glands to where they are needed	In the plasma
6 Heat from the liver and muscles to the rest of the body so it is all at the same temperature	In all parts of the blood

Blood protects us

1 White cells called **phagocytes** eat germs.

2 Other white cells make **antibodies** to fight disease. Antibodies destroy the germs that cause disease. Antibodies also destroy toxins that germs make. There is a different antibody for each disease.

Once your white cells have made a particular kind of antibody, they can make it faster next time. Also it may stay in your blood for a while. This makes you **immune** to the disease. You will not catch it again, or else you will have only a mild attack.

You can get **vaccinated** against some diseases. Specially treated germs are injected into you, to give you a mild attack of the disease. Your body makes antibodies, so you become immune for the future.

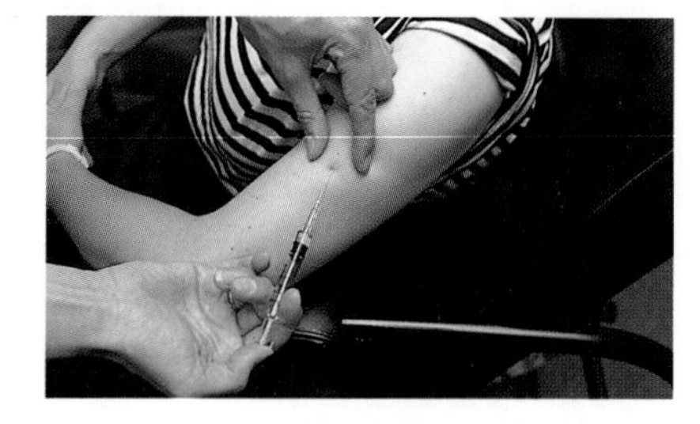

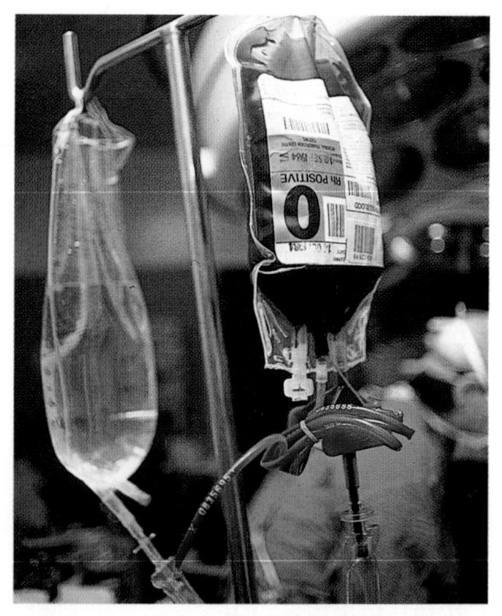

Often, ill people need blood transfusions. The blood comes from donors. It is sent out to hospitals in plastic packs.

Questions

1 What part of the blood:
  a) carries digested food?
  b) helps protect you against disease?

2 Describe two ways in which your body can become immune to a disease.

3 What do antibodies do?

5·6 How oxygen and food reach cells

Blood carries oxygen and food around the body. But it never comes into contact with body cells. So how does food and oxygen get from the blood to the cells where they are needed?

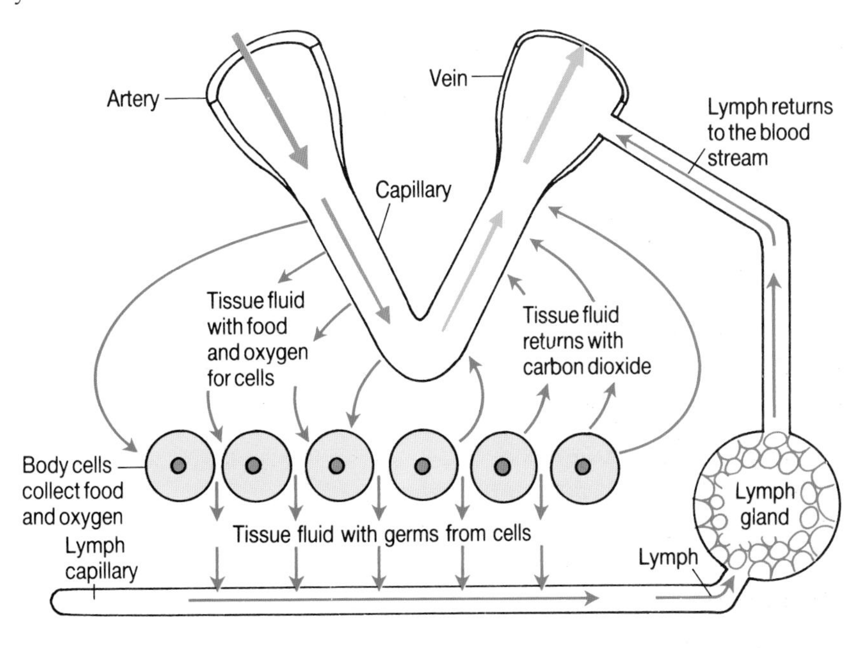

Tissue fluid

The answer is that capillary walls are so thin that they leak. A liquid called **tissue fluid** leaks through them from the blood, into tiny spaces between the body cells.

Tissue fluid carries oxygen and food from the blood to the cells, and washes away their wastes.

Most tissue fluid then seeps back into the blood vessels. But some drains into **lymph capillaries**, carrying germs and bits of dead cells with it. It becomes **lymph**.

Where does the lymph go?

The lymph capillaries join to make larger tubes. These tubes drain into **lymph glands**.

Here the lymph is cleaned up by white cells called **lymphocytes**. These eat the germs and dead cells, and also make antibodies.

Clean lymph is returned to the blood stream through tubes which join a vein in the neck.

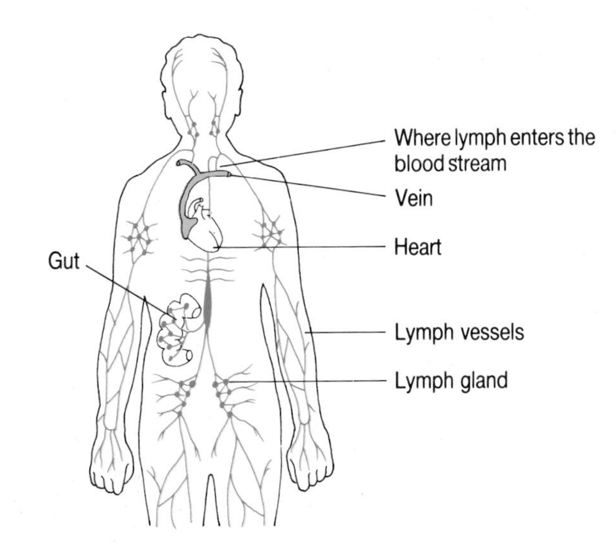

Questions

1 How does food and oxygen reach your body cells?
2 What happens to tissue fluid after it has been to the cells?
3 What happens inside lymph glands?
4 How does lymph get back into the blood stream?
5 Where in your body are the lymph glands?

5·7 Skin and temperature

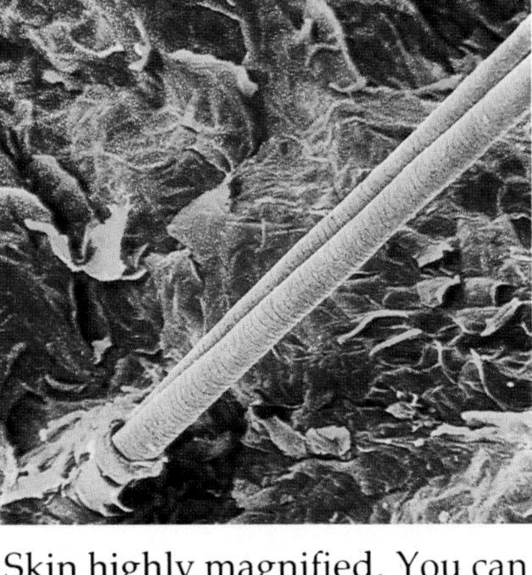

Skin highly magnified. You can see the dead cells flaking off. What do you think the spike is?

What skin does

1 It **protects** your body against damage, dirt, and germs.

2 It contains millions of tiny **sense organs**, which are sensitive to touch, temperature, and pain.

3 It **excretes** water and salts from your body, as **sweat**.

4 It helps keep your body at a **steady temperature**.

The structure of skin

This is a drawing of a tiny piece of skin, greatly magnified.

The **outer layer** of your skin is a tough protective layer made of dead, flat cells. Dead cells continually flake off. But they are replaced by cells from below, which grow and flatten out as they move to the surface.

Sebaceous glands produce an oily substance called **sebum**. This makes skin and hair supple and waterproof. It also slows the growth of germs.

The **hair root** is the only living part of a hair. It is deep inside the skin in a tube – the **hair follicle**. Hair protects your head from direct sunlight and, in hairy animals, keeps the body warm.

Hair

Pore

Capillaries near surface of the skin

Blood vessel

This layer of cells is full of **fat and oil**. It helps keep the body warm.

Sweat glands produce liquid called sweat, to cool the body when it gets too hot.

Skin colour

Some skin cells make melanin, the brown pigment responsible for skin colour. Skin makes more melanin in sunshine, because melanin protects it from the sun's harmful ultraviolet rays.

Dark skin contains more melanin than pale skin does.

Skin and body temperature

The outside of your body may feel hot or cold. But inside, it always stays at about the same temperature, 37°C. Your skin helps to keep it at this temperature.

But your skin alone won't keep you cool in hot weather. . .

When you get too hot:

1 You sweat a lot. The sweat evaporates and cools you down.

2 The blood vessels below your skin **expand**. So a lot of blood flows near your body surface and loses heat – like a radiator.

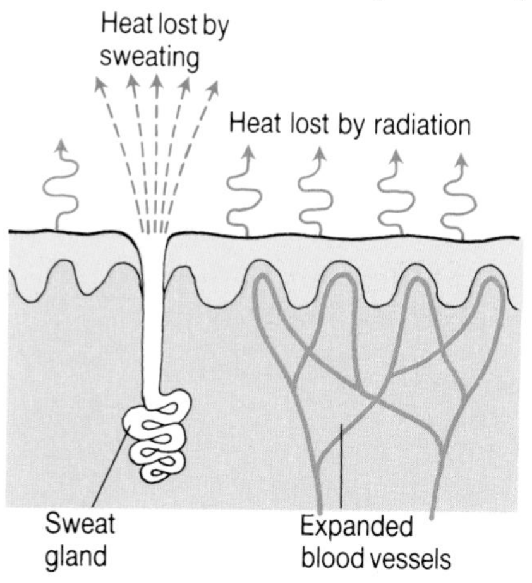

When you get too cold:

1 You stop sweating.

2 Blood vessels below the skin **contract**. So only a little blood flows near the surface and loses heat.

3 Muscles start rapid, jerky movements we call **shivering**. This produces extra heat, which warms up the body.

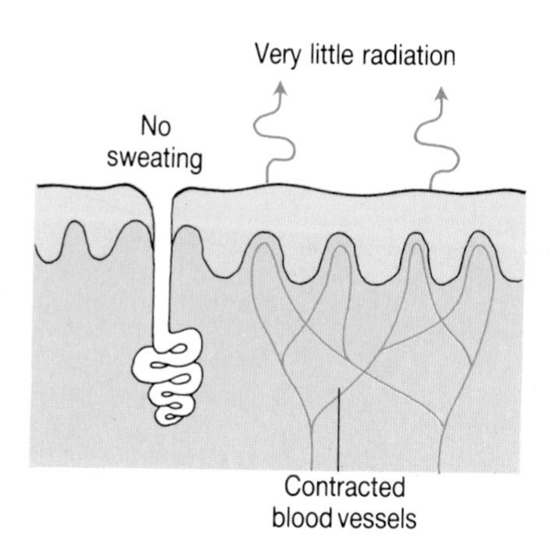

. . . or warm in cold weather. You help it out by wearing the right kind of clothes.

Body hair

Many animals have a thick coat of hair, or fur. This traps a layer of warm air around the body.

In cold weather, tiny muscles make the hairs stand up. So more air is trapped. In warm weather the hair muscles relax. So less air is trapped.

A seal pup, all dressed up for arctic life.

Questions

1 Your skin is being worn away all the time. How is it replaced?

2 How is skin protected by a suntan?

3 What changes happen in your skin when you get too hot?

4 What changes happen in your skin when you go from a warm room out into a cold winter night?

5 How does fur help artic animals to keep warm?

5·8 Excretion

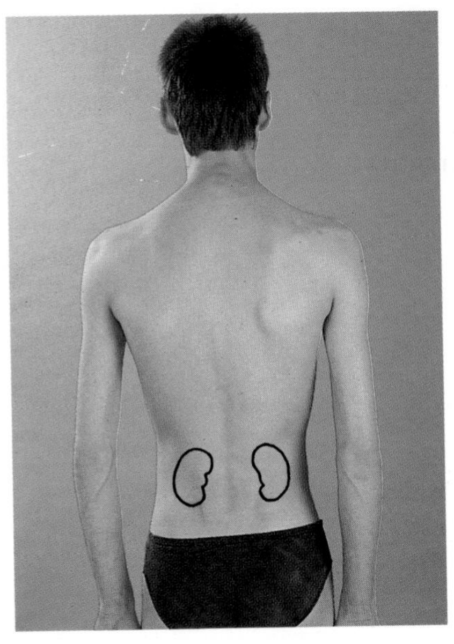

Where your kidneys are . . .

All day long, even while you sleep, your body produces waste substances. These wastes include carbon dioxide and urea.

Your body must get rid of these things as they are poisonous. Waste removal is called **excretion**. Your lungs excrete carbon dioxide. Your kidneys excrete urea. Your kidneys are your main **excretory organs** – they are towards the back of your body just above your waist.

What kidneys do

Your kidneys remove urea, water, and other unwanted substances from your blood. Urea is a waste produced by your liver.

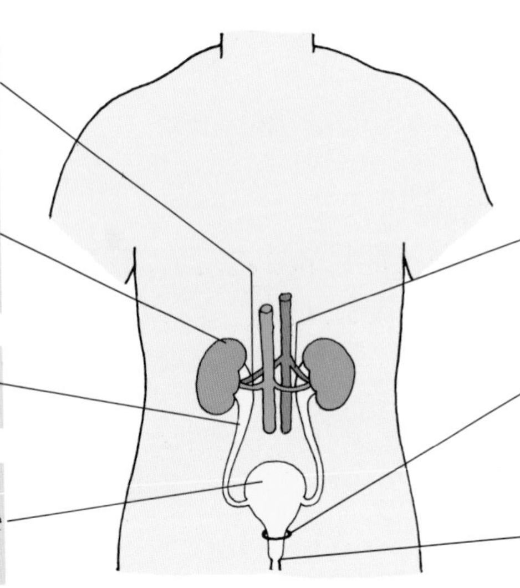

The **renal vein** carries 'clean' blood away from the kidneys.

The **kidneys** remove urea and other wastes from blood, and excrete it in a liquid called **urine**.

Ureters are tubes which carry urine to the bladder.

The **bladder** is a bag which stores urine until you go to the toilet.

The **renal artery** carries 'dirty' blood to the kidneys.

The **sphincter** is a ring of muscle which keeps the bladder closed until you go to the toilet.

The **urethra** is a tube which carries urine out of your body.

How kidneys clean blood

Kidneys clean blood by filtering it. They filter all your blood 300 times a day. The filtering is done by over a million tiny tubes packed into each kidney. These tubes are called **nephrons**.

Some people's kidneys are not very good at filtering blood. Kidney machines help by filtering their blood for them. The little girl in the photograph is attached to a kidney machine.

Surgeons can sometimes replace faulty kidneys with healthy ones. This operation is called a **kidney transplant**.

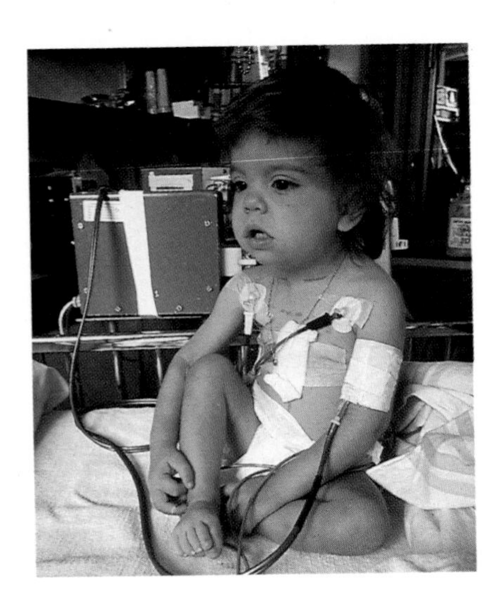

The parts of a kidney

This diagram shows a kidney cut in half, with one nephron magnified.

A **nephron** begins with a cup-shaped bag containing a bunch of capillaries called a **glomerulus**. 'Dirty' blood enters a glomerulus and is filtered.

Filtered liquid enters this tube-shaped part of a nephron. Here it is turned into **urine**.

The urine from many nephrons drains into a **collecting duct**.

The urine drains into the **ureter**.

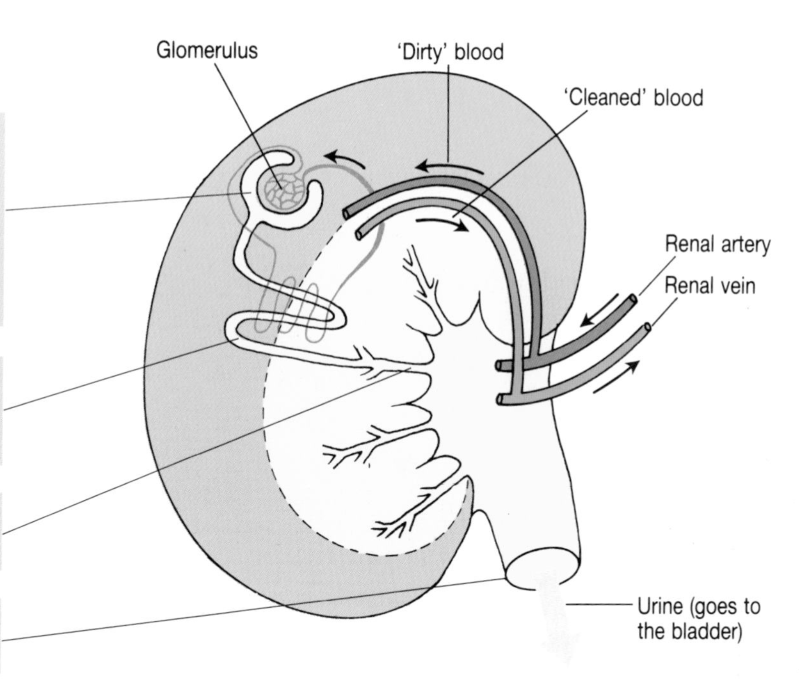

How a nephron works

The diagram below shows one nephron straightened out.

Blood is **filtered** by the glomerulus. Nearly all the blood except red cells filters through into the nephron.

The liquid in the nephron contains useful substances like glucose and vitamins as well as urea. These useful substances pass back into the blood – they are **reabsorbed**.

The liquid left in the nephron is **urine**. It contains urea and water plus other unwanted substances. It goes through the collecting ducts and ureters to the bladder.

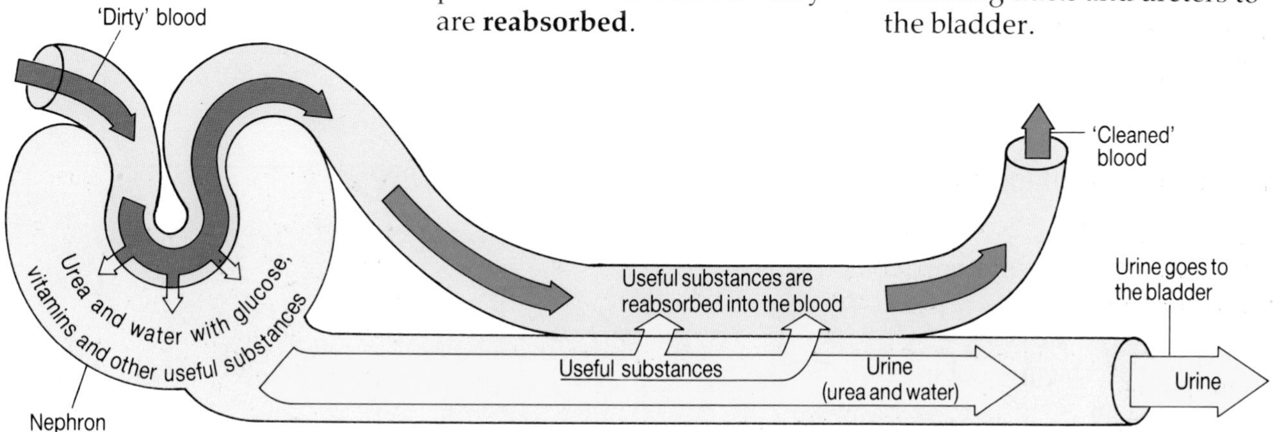

Questions

1 Why would you die if excretion stopped?

2 What do lungs and kidneys excrete?

3 What do the bladder and sphincter do?

4 Where is 'dirty' blood filtered?

5 What happens to stop useful substances being lost from the body as urine is produced?

Questions

1 This diagram shows the inside of the heart. Match the labels on the diagram with the descriptions below:

a) right atrium;
b) blood on its way around the body;
c) valve tendon;
d) right ventricle;
e) a vein carrying blood from around the body;
f) left ventricle;
g) valve which stops blood flowing back into the left atrium;
h) cardiac muscle;
i) blood on its way to the lungs;
j) blood returning from the lungs;
k) valve which stops blood flowing into the left ventricle.

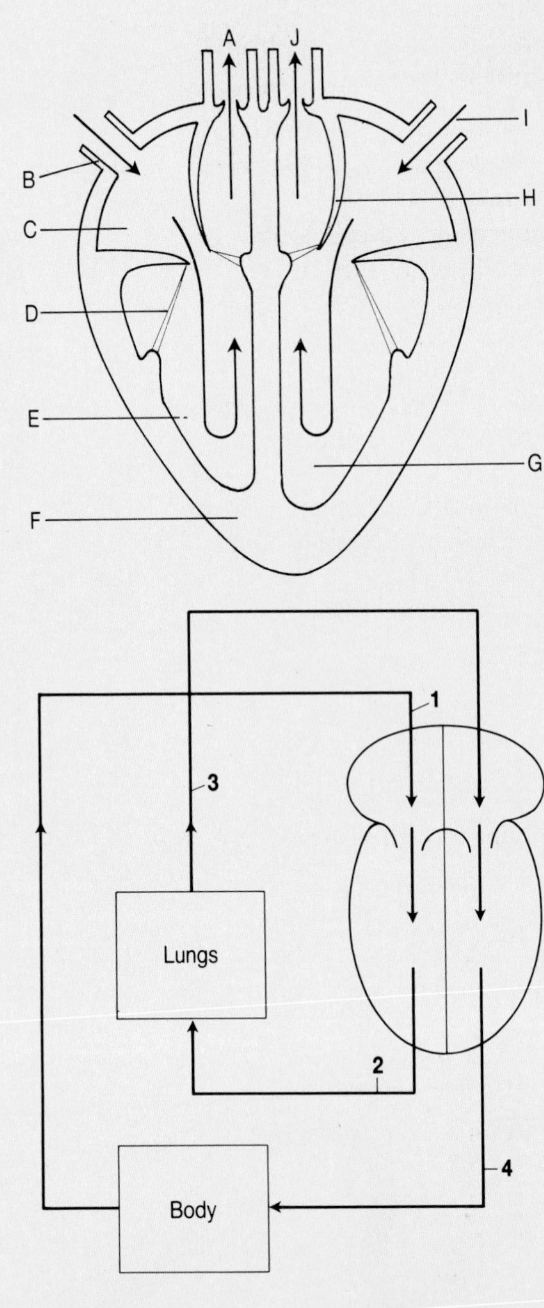

2 Opposite is a simpler version of the diagram from page 79, showing the heart and main blood vessels.

Which of the blood vessels labelled 1 to 4:

a) contains blood at the highest pressure in the whole body?
b) contains blood at very low pressure?
c) contains blood full of carbon dioxide?
d) contains blood full of oxygen?

3 a) Explain what these medical words mean: angina, thrombosis, heart attack.
b) Why is it important not to eat too much fatty food?
c) Name ten foods which contain lots of fat.
d) Apart from fatty food, what else can cause heart disease?
e) List all the things you should do to avoid heart disease.

4 This is a drawing of human blood cells.

a) Name each cell.
b) Which of the cells:
contains haemoglobin?
is a phagocyte?
can pass through capillary walls?
transports oxygen?
eats germs?
gives blood its red colour?
protects the body?

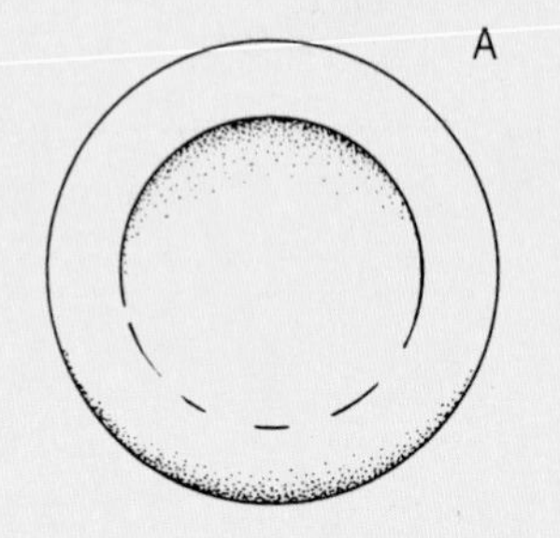

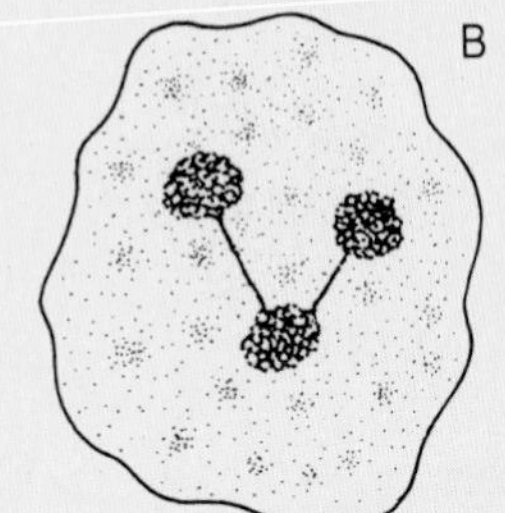

5 This diagram shows parts of the human skin:
a) Name the parts labelled A to H.
b) Name the parts which help to control body temperature.
c) What do these parts do when we get too hot?
d) What do these parts do when we get too cold?
e) Explain how part B is replaced as fast as it wears away.
f) Why does layer H grow thicker in wild animals in autumn?

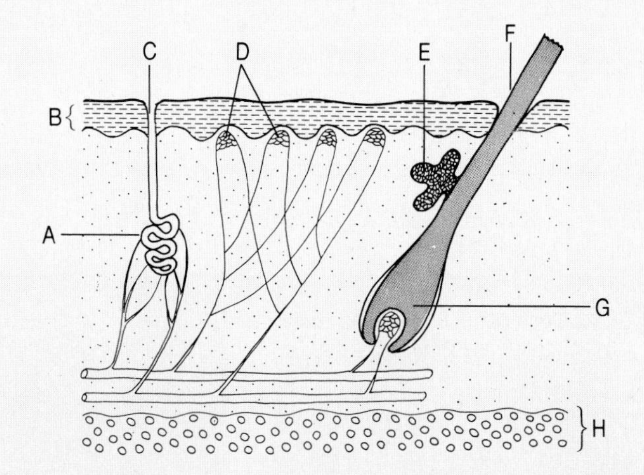

Investigations

1 The heart beat
a) What is your pulse?
b) Find your pulse using the method described on page 79.
c) Count how many times your pulse beats a minute while you are relaxing.
d) Now run on the spot for 30 seconds. Stop, and immediately count how many times a minute your pulse is beating.
e) How does your pulse change during exercise?
f) Name two things your muscles need more of during exercise.
g) How does this explain what happens to your pulse during exercise?

2 Invertebrate heart beat
Under a microscope you can see the heart beating inside a water-flea (*Daphnia*). You can find water-fleas in ponds and streams in spring and summer.
a) Put one water-flea in water on a cavity microscope slide, and look at it through a microscope. Observe its heart beating.
b) Its heart beats very fast. But you can count the beats by tapping the point of a pencil on paper in time with them, then counting the pencil marks.
c) Use this method to find out how many times its heart beats in five seconds:
when it is in water at room temperature;
when the water has been cooled in a fridge.
Ask a friend to tell you when the five seconds are up.
d) When an animal is warm it uses energy faster than when it is cool. How does this explain your results?

3 Looking at the capillaries in a fish tail
You can see blood flowing through capillary blood vessels in the tail of a goldfish, under a microscope.
a) Wrap a goldfish in wet cotton wool, taking care not to damage its scales.
b) Put it in a small dish and cover its tail with half a microscope slide.
c) Look at the thin part of the tail, under high magnification.
Do not keep the fish under these conditions for more than ten minutes.

4 Why you sweat when you are hot
Sweating cools you down because the sweat takes heat away from your skin when it evaporates. You can feel it happening in this experiment:
a) Using a wet tissue, make a wet patch on the back of one hand. Blow gently on it. The wet patch feels cooler than the rest of your hand because the water is evaporating.
b) If you use alcohol instead of water the wet patch feels even cooler, why do you think this is?
c) Why does wet skin feel cooler when you blow on it?
b) How does this investigation explain why you feel cooler in hot windy weather than in hot still air?

6·1 Food

Why we need food

For growth and repair. Your body grows by forming new cells. You also need new cells to replace dead ones. Cells are built from substances in food.

For energy. You need energy to work your muscles, and all other organs. The energy in food is measured in **calories** or in kilojoules.
(4.2 kilojoules = 1 Calorie)

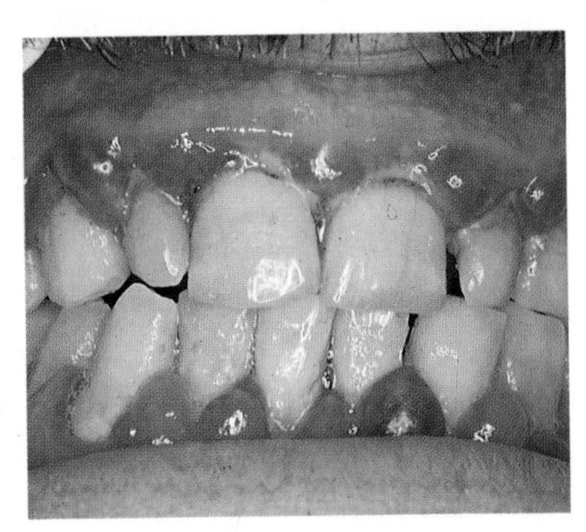

To stay healthy. You need vitamins and minerals in food. This shows what happens to gums if you don't get enough vitamin C.

What's in food

Food contains a mixture of substances. The main ones are proteins, carbohydrates, fats, oils, vitamins, minerals, fibre, and water.

These eggs contain:

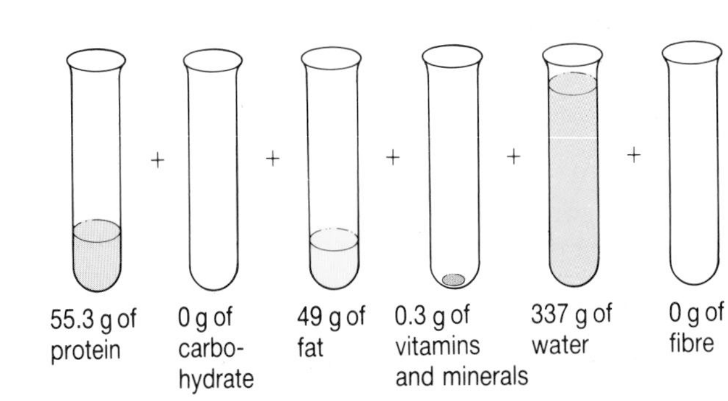

These beans contain:

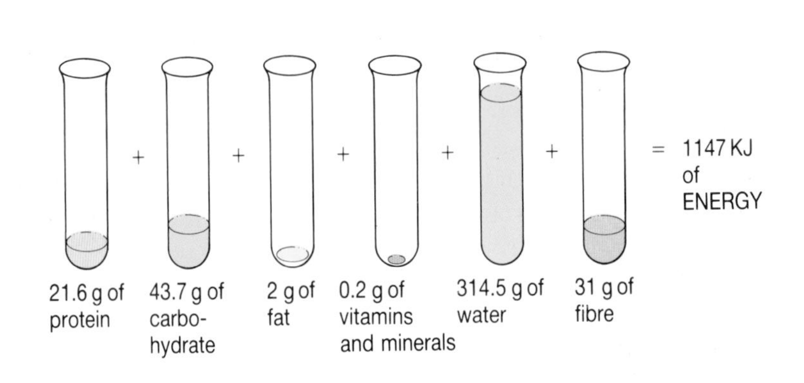

What foods do for you

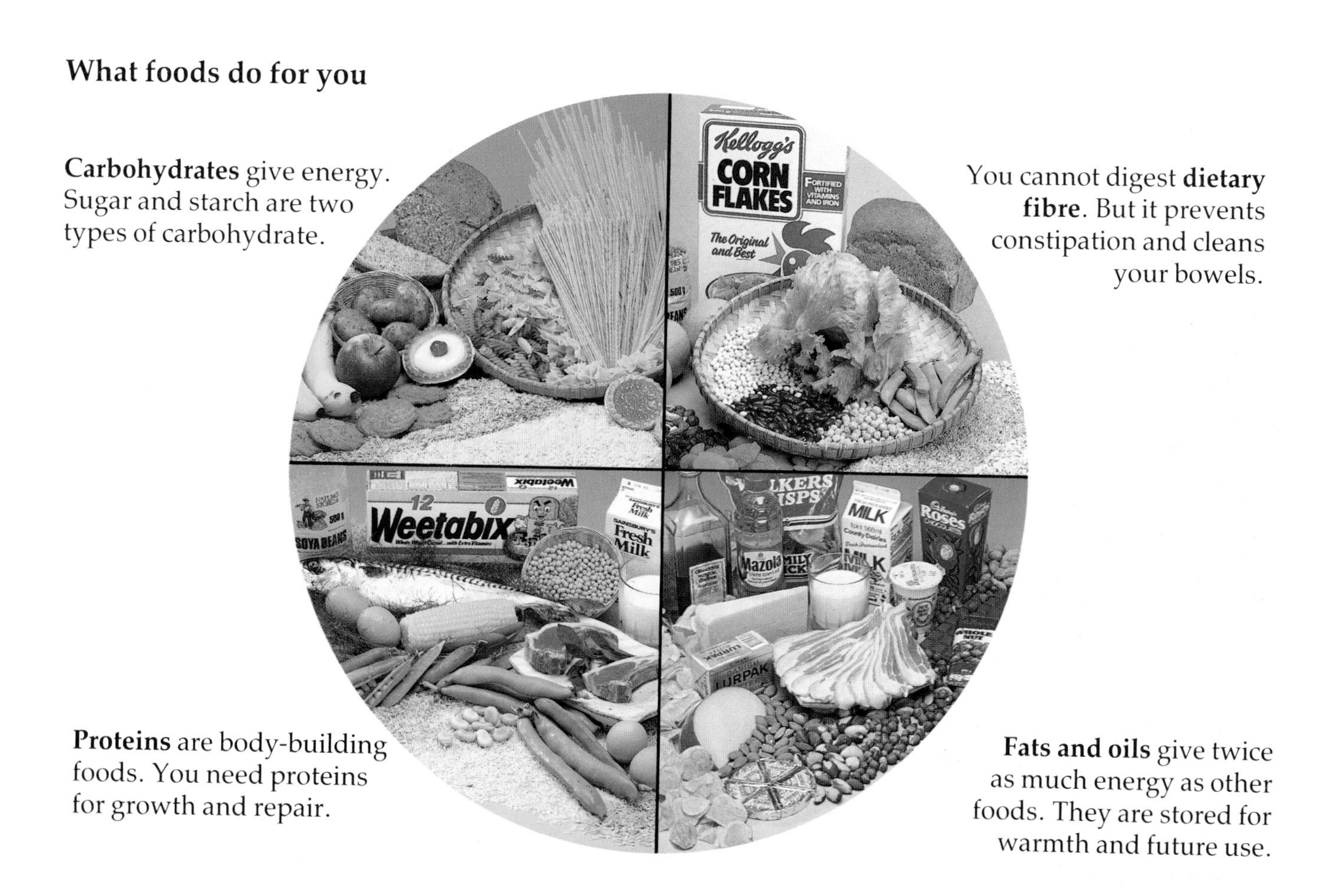

Carbohydrates give energy. Sugar and starch are two types of carbohydrate.

You cannot digest **dietary fibre**. But it prevents constipation and cleans your bowels.

Proteins are body-building foods. You need proteins for growth and repair.

Fats and oils give twice as much energy as other foods. They are stored for warmth and future use.

The main vitamins and minerals you need

Substance	Where you find it	Why you need it	Shortage can cause
Vitamin C	Oranges, lemons, grapefruit, green vegetables, potatoes	For healthy skin and gums and to heal wounds quickly	**Scurvy.** Gums and nose bleed. The body bleeds inside
Vitamin D	Milk, butter, eggs, fish, liver (also made by skin in sunshine)	For strong bones and teeth	**Rickets.** The bones become soft and bend
Calcium	Milk, eggs	For strong bones and teeth	**Rickets.**
Iron	Liver, spinach	For making red blood cells	**Anaemia.** The person is pale and has no energy

Questions

1 Name three foods rich in:
- **a)** dietary fibre
- **b)** sugar
- **c)** protein
- **d)** starch
- **e)** fats and oils.

2 Name foods needed:
- **a)** for strong bones and teeth
- **b)** to avoid scurvy
- **c)** to avoid constipation
- **d)** for warmth
- **e)** for growth.

6·2 Eating for health

If you ate only sweets, or cream buns, or crisps you would stay alive for a time. But you would not stay healthy, because you are not eating a balanced diet.

A **balanced diet** is one which contains the right amounts of protein, carbohydrate, fat, vitamins, and minerals to satisfy all your body's needs.

How much should you eat?

The amount of food you need depends on the energy you use in a day. And that depends on:

1 your age. Teenagers use more energy than babies

2 your work. A footballer uses more energy than a snooker player

3 your sex. Males use more energy than females of the same age, even for the same work

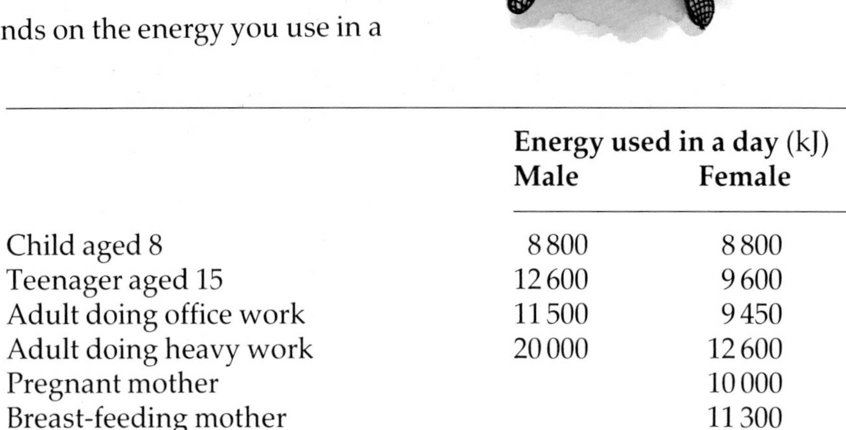

	Energy used in a day (kJ)	
	Male	**Female**
Child aged 8	8 800	8 800
Teenager aged 15	12 600	9 600
Adult doing office work	11 500	9 450
Adult doing heavy work	20 000	12 600
Pregnant mother		10 000
Breast-feeding mother		11 300

This table shows the amount of energy used in a day.
Energy used by the body is measured in kilojoules (kJ for short).

Getting the balance right

The food you eat each day should supply just enough energy to get you through that day.

If you eat too much, your body stores the extra as fat. So you get overweight, or **obese**. Obese people are more likely to have heart attacks than slim people.

If you eat too little, you lose weight. You feel weak and have no energy. Some girls eat so little that they suffer from **anorexia**. This happens if a girl starts a slimming diet and doesn't know when to stop. But we have to eat to live – so get the balance right.

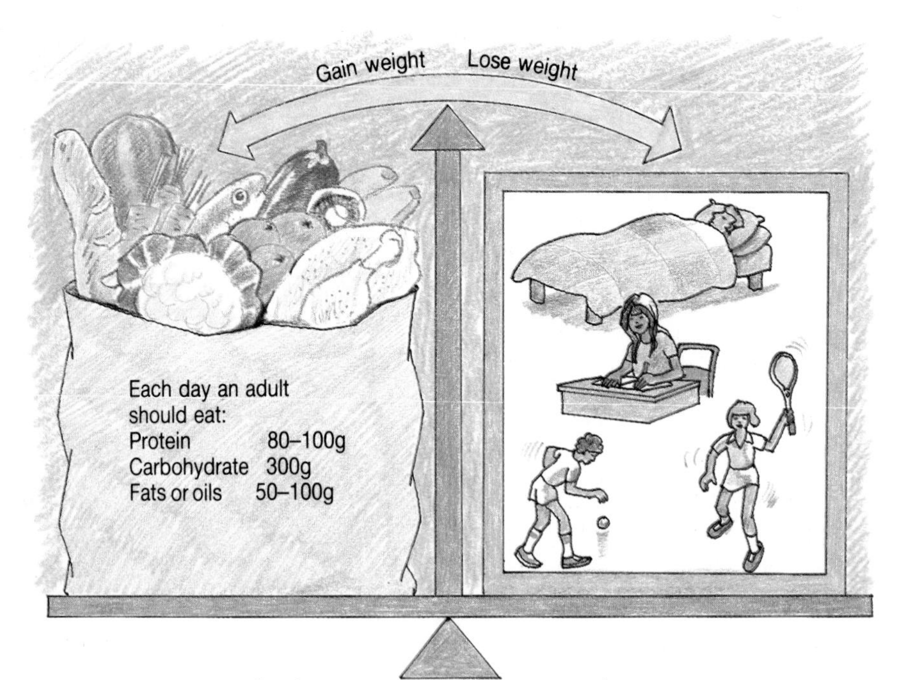

Food eaten must **balance** energy used

Healthy eating

Some foods are bad for you if you eat too much of them. So it makes sense to eat healthy foods, and avoid those which can damage your health.

A little fat gives you a *lot* of energy. So it is easy to eat more than you need. Fats make you fat, and can cause heart disease.

Sugar gives you energy but nothing else. It rots your teeth and also makes you fat. Cakes, chocolate, and ice cream also contain hidden fat.

During processing in factories, food loses dietary fibre and other goodness. Salt, sugar, colouring, and other chemicals are added. These may do you harm.

These foods contain protein but not much fat. They build you up without making you fat.

These give you carbohydrate, protein and fibre. They fill you up without making you fat.

These foods contain dietary fibre, vitamins, and minerals to keep you healthy.

Questions

1. Name the five types of food which make up a balanced diet.
2. Why does a teenager need more food than a baby?
3. What happens if you eat more than you need?
4. What happens if you eat less than you need?
5. Explain why you should:
   a) eat less cakes and ice cream
   b) cut most of the fat from meat
   c) eat fresh food instead of processed food
   d) eat more fruit and vegetables.
6. Name five foods that contain fibre.

6·3 Food additives

Yum – additives!

When food is processed in factories, chemicals called **food additives** are often added.

Around 3500 additives are used by British food manufacturers. Below are the main kinds.

Colourings make food look more appetizing. Tinned peas and strawberries owe their bright colours to added chemicals. Egg yolks are given a brighter colour by chemicals put in hen food.

Flavourings put back flavours lost when food is processed in factories. They are sometimes used instead of 'real' flavours. Fruit-flavoured puddings and drinks owe their taste to chemicals and not real fruit.

Preservatives help to keep food fresh. So food can be stored in shops for some time before it goes bad. Pork pies, sausages, bacon, and dried fruits have preservatives in them.

Are additives dangerous?

There is no proof that additives cause serious illness in humans. But here are a few reasons why some people avoid eating these chemicals:

1 Additives fed to animals have caused damage to their kidneys, liver, and digestive system, and some have caused cancers.

2 Some additives destroy vitamins in food. For example, the preservative sulphur dioxide destroys vitamin B_1.

3 Some may cause asthma, itchy skin, headaches, and **hyperactivity** (uncontrollable behaviour) in young children.

If you want food free of additives, eat fresh vegetables and fresh fruit, fresh meat, real-fruit juices and yoghurt, and muesli.

Black Forest Gateau

INGREDIENTS

Sponge: Wheat flour, sugar, liquid egg, water, cocoa powder, dried skimmed milk, glucose, emulsifiers E471 E475, starch, salt, flavouring, colours E102 E122 E124 E142 155. Filling and Decoration: Cream (cream, dextrose, stabiliser E407), cherry topping (cherries, sugar, glucose, modified starch E1422, citric acid E330, flavouring, colours E123 E142 E422, thickener E412, preservative E211), chocolate flavoured coating (sugar, vegetable fat, fat reduced cocoa powder.

How many additives are there in this gateau? An additive is shown by the letter E, followed by a number.

Questions

1 Why are additives put in food?

2 Why do some people avoid food that contains additives?

3 Have a competition to see who can find the food containing the most additives. (Look on the labels of food packets and tins)

6·4 Teeth

Teeth grow out of holes called **sockets** in your jaw bone. Before digestion starts teeth are used for biting and chewing food.

Enamel forms the hard, biting surface of a tooth.

Dentine is similar to bone.

The **pulp cavity** contains nerve endings and blood vessels.

Cement holds the tooth in its socket.

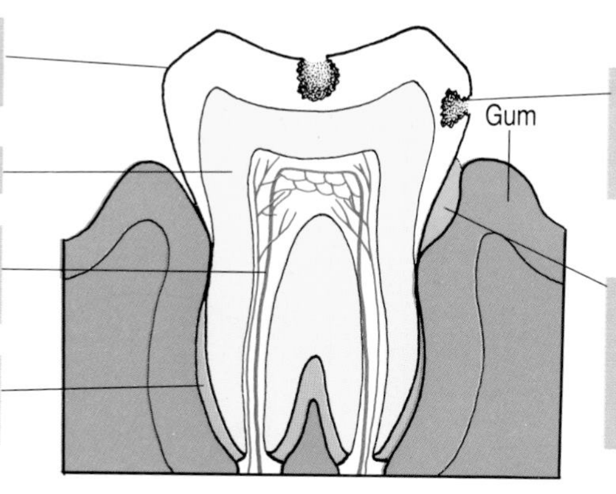

Tooth decay starts on the top of large back teeth, and where one tooth touches another.

Gum disease starts between the gum and teeth. It can destroy tooth cement and cause teeth to fall out.

Why teeth decay

Tooth decay happens when the **bacteria** in your mouth turn the **sugar** in your food into **acid**. The acid eats a hole in tooth enamel and dentine. When the hole reaches a nerve you get toothache.

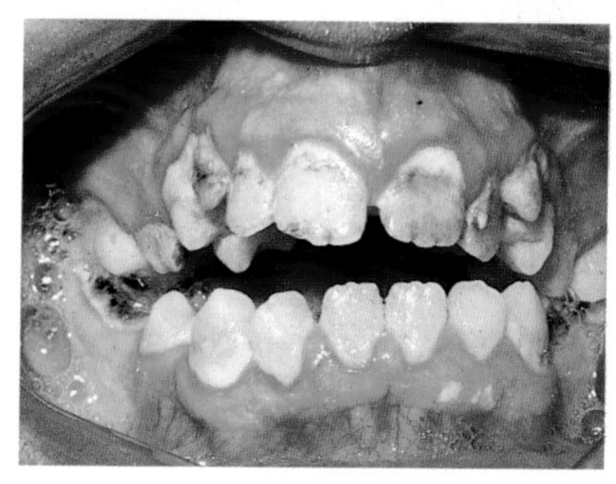

. . . so don't forget to brush your teeth!

Gum disease

If you don't clean your teeth regularly they become covered with a layer of food and bacteria called **plaque**. Gum disease starts if the plaque gets between the gum and a tooth. The plaque rots and causes the gum to swell and bleed.

Preventing tooth decay and gum disease

1 Eat less sugary foods, because **sugar** + **bacteria** = **acid** which rots teeth.

2 Brush your teeth at least once a day to remove plaque.

3 Use fluoride toothpaste. Fluoride strengthens tooth enamel.

4 Visit a dentist every six months.

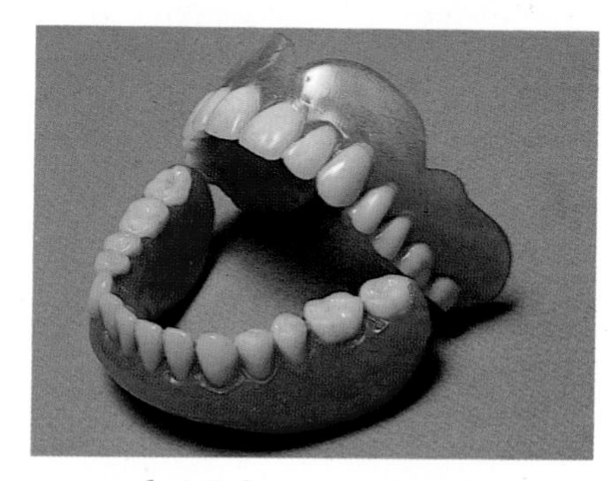

. . . so don't forget to brush your teeth!!

Questions

1 What is plaque?

2 How does plaque cause tooth decay and gum disease?

3 Why should you use fluoride tooth paste?

4 Name three foods which can rot your teeth. Explain why in each case.

6·5 Digestion

Bread and bacon and rice and all the other things you eat must be changed into a liquid inside your body, before your body can use them. The breaking down of food into a liquid is called **digestion**.

Food is digested inside a tube in your body called the gut, or **alimentary canal**. Your alimentary canal is over seven metres long and coiled inside your body.

No matter what you eat, it is turned into liquid in your gut.

Where your food is digested

Digestion starts when food is taken into the mouth. This is called **ingestion**. The teeth are used to break the food into small pieces. The pieces are mixed with a liquid called **saliva** and swallowed.

The walls of the gut produce chemicals called **digestive enzymes**. Enzymes break down food into liquids.

Liquid food passes through the gut wall into the blood stream. This is called **absorption**. Blood carries food to all parts of the body.

Food is taken in by cells, and used for energy, growth, and repair. This is called **assimilation**.

Dietary fibre and other things in food which cannot be digested pass out through the anus when you visit the toilet. This is called **defecation**.

Substances which cannot be digested are called **faeces**.

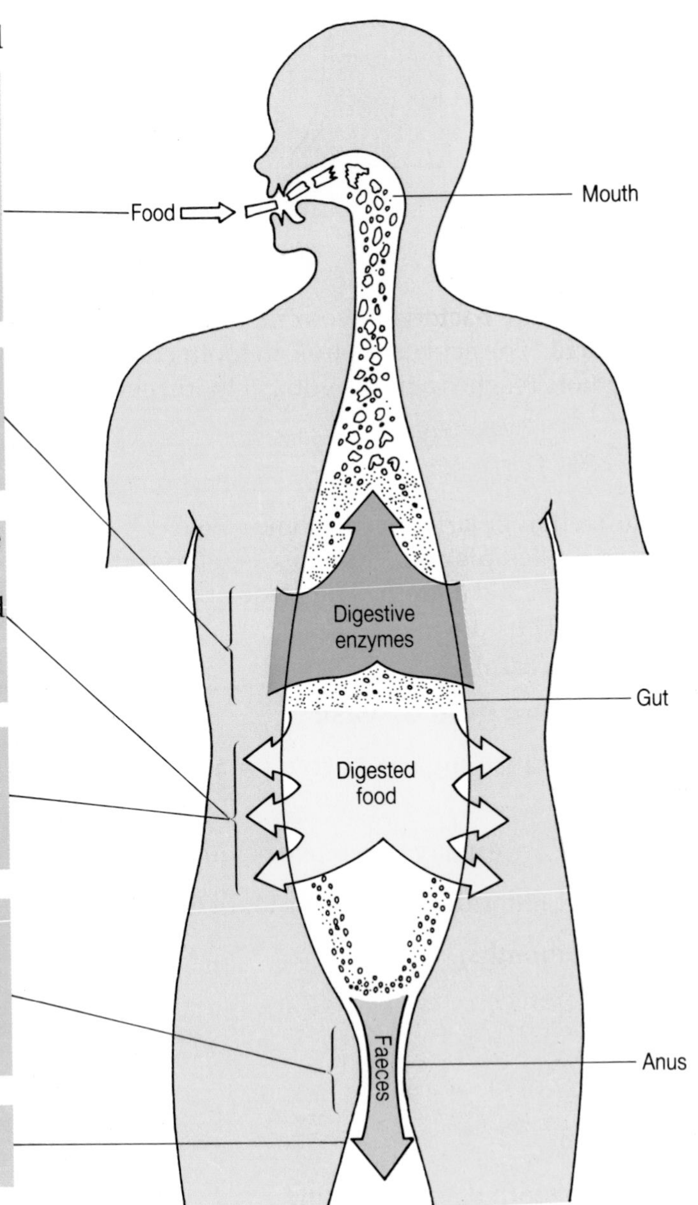

More about digestive enzymes

Digestive enzymes change the food you eat by breaking down large food molecules into smaller ones which can be absorbed into the blood. Each type of food needs a different enzyme to break it down.

Amylase enzymes break down carbohydrates like starch into glucose.

Lipase enzymes break down fats and oils into fatty acids and glycerol.

Protease enzymes break down proteins into amino acids.

This is what happens when starch is broken down by amylase enzyme and absorbed.

Starch is a carbohydrate found in rice, bread, and potatoes. Starch molecules are made up of many glucose molecules joined together, like a string of beads.

Amylase enzymes break down the starch by cutting it up into separate glucose molecules, like you might cut up a string of beads with scissors.

Glucose molecules are so small that they can pass through the cells which form the gut wall, and then through blood vessel walls into the blood stream.

Blood carries the glucose to the cells of the body for assimilation.

Other kinds of food are broken down by other enzymes in the same way.

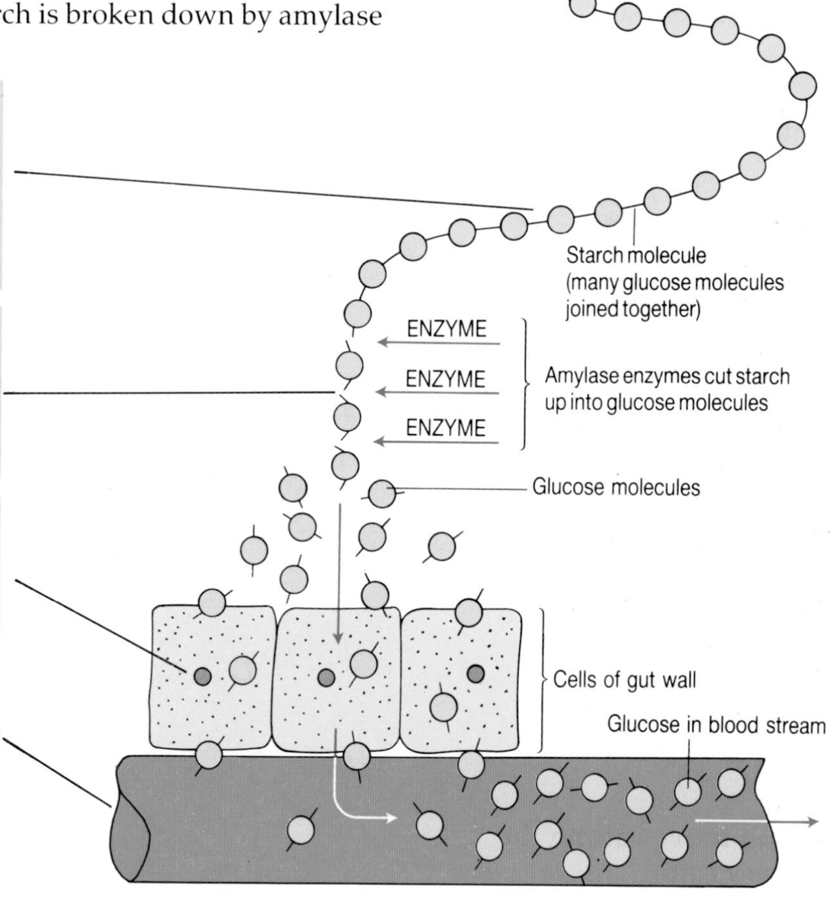

Questions

1 What is your alimentary canal?

2 What is the scientific word for:
  a) taking food into your mouth?
  b) breaking down food into small molecules?
  c) movement of food molecules into the blood?
  d) getting rid of undigested food?

3 Name the enzymes which digest: protein, carbohydrates, fats.

4 Which of the foods in question 3 are digested into: amino acids? glucose? fatty acids and glycerol?

6·6 A closer look at digestion

Digestion in the mouth

Digestion starts in your mouth. When you chew, food is ground into a pulp. At the same time, it is mixed with **saliva** from salivary glands. Saliva does two main things:

1 It wets the food, so that it slips easily down your throat.

2 It contains an **amylase enzyme** which starts to digest the starch in your food into sugar.

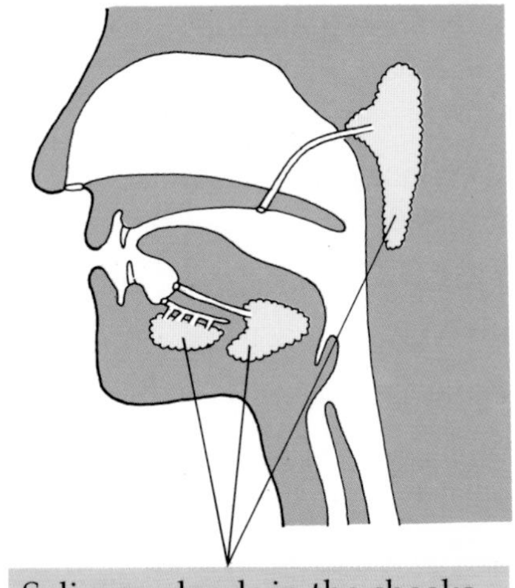

Salivary glands in the cheeks and under the tongue.

Swallowing

Before you swallow food, your tongue shapes it into a round lump called a **bolus**, and squeezes it to the back of your mouth.

The bolus pushes the **soft palate** upwards. This stops food getting into the space behind your nose.

A flap of skin called the **epiglottis** drops over the top of the wind-pipe. This stops food getting into your lungs.

The bolus is squeezed past the epiglottis, into your gullet or **oesophagus**.

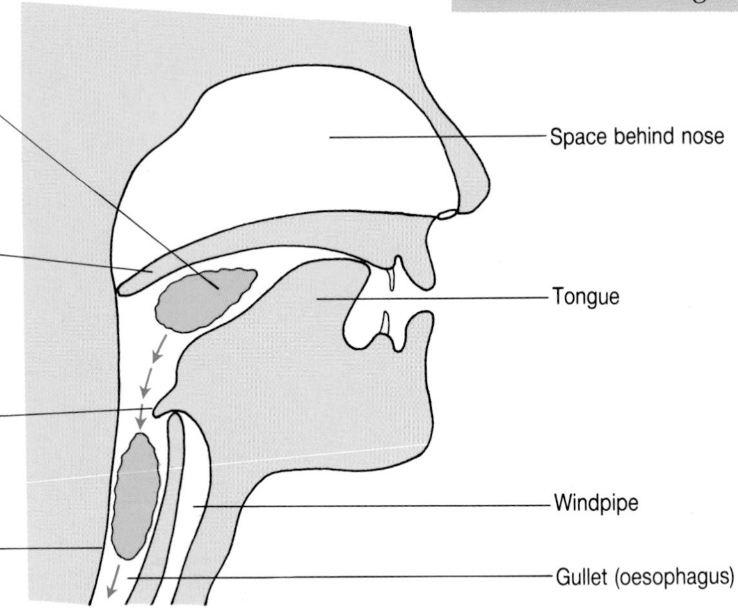

How food moves along the alimentary canal

Your gullet, and the rest of your alimentary canal, has **circular muscles** in its walls.

These muscles contract behind a bolus, and relax in front of it. So the bolus gets pushed along.

This contraction and relaxation of circular muscles is called **peristalsis**. It takes place all the way along your alimentary canal, and keeps food moving.

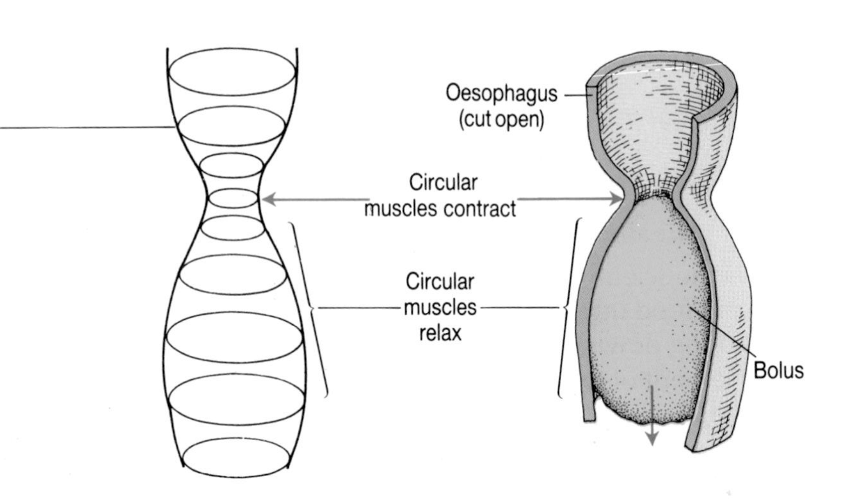

Digestion in the stomach and intestine

This is a simplified diagram of the alimentary canal. The drawing on page 102 shows what it really looks like, and its position in the body.

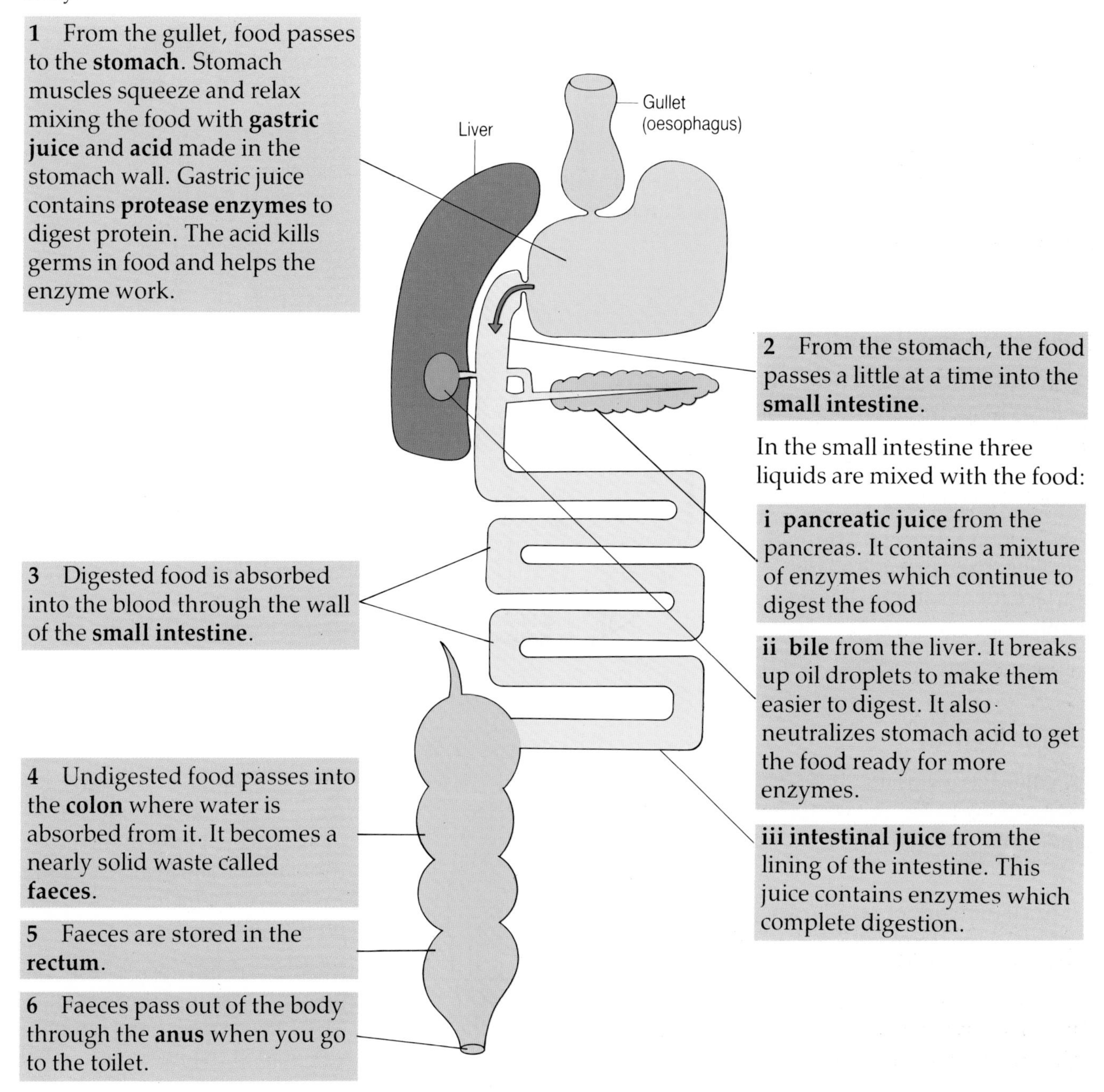

Questions

1. Where in your body does digestion begin?
2. What are salivary glands for?
3. What might happen if you had no epiglottis?
4. What is the scientific name for your gut?
5. What substances are mixed with food in your stomach? How does each help digestion?

6·7 Absorption, and the liver

Digestion breaks down food so it can be absorbed into your blood and carried to all parts of your body. Carbohydrates are digested into glucose, proteins into amino acids, and fats and oils into fatty acids and glycerol. But vitamins, minerals, and water in food don't have to be digested since they are already easily absorbed.

Digested foods are absorbed through your gut wall into the blood and carried away. Most absorption goes on in your small intestine.

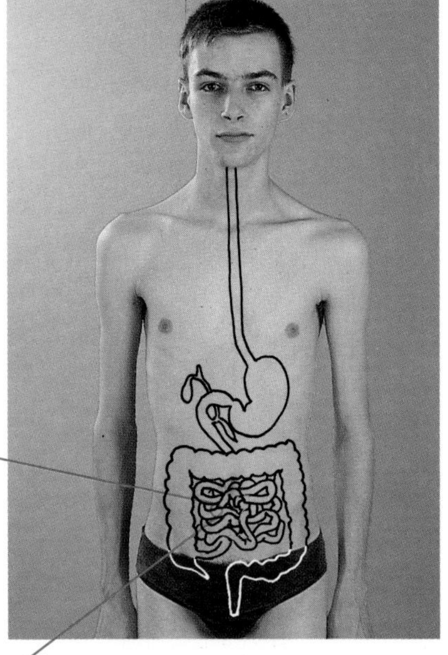

Where your small intestine is . . . It extends down from your waist.

Inside the small intestine

Your small intestine is over six metres long. Inside, millions of tiny fingers called **villi** stick out of its walls. One villus is about a millimetre long.

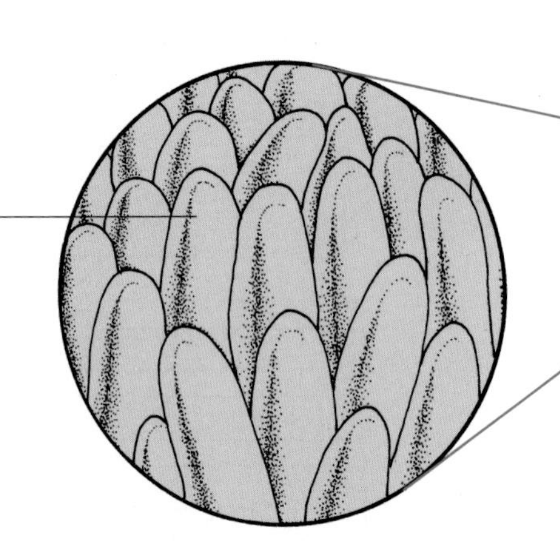

Absorption takes place through the villi. They provide a large surface for absorbing digested food.

Inside a villus

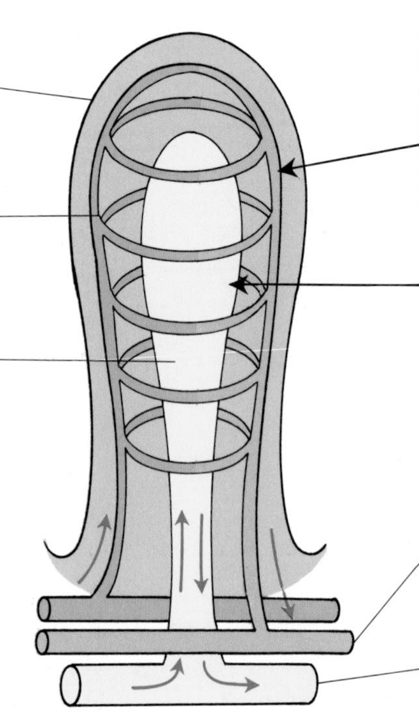

A villus has a surface layer only one cell thick. This lets digested food pass through easily.

It has a network of **capillaries** which carry blood.

It has a lymph vessel called a **lacteal** which carries lymph.

Glucose, amino acids, and some fatty acids and glycerol pass into the blood in the capillaries. Blood carries them to the liver, which is described on the next page.

Most of the fatty acids and glycerol pass into the lacteal. They are carried in the lymph to a vein in the neck. From there they flow into the blood and are carried to all parts of the body.

Blood vessel carrying digested food to the liver.

Lymph vessel carrying digested food to the blood stream.

What happens to digested food in the liver

Your liver is the largest organ in your body. It weighs over one kilogram and stretches all the way across your body, just above the waist.

The liver is a chemical factory, a food store, and a central heating system. Here are just a few of the jobs it does:

1 It stores glucose as **glycogen**. It changes this back to glucose when the body needs it.

2 It stores the minerals copper and potassium, as well as iron needed to make red blood cells.

3 It stores vitamins A, B, and D.

4 It takes the goodness out of unwanted amino acids, and changes what is left into a waste called **urea**. Urea is removed from your body by your kidneys.

5 It takes some poisons from the blood and makes them harmless. These poisons come from germs, alcohol, and drugs.

6 It makes **bile**, which is needed for digestion.

7 It makes **fibrinogen**, which is needed for blood to clot in wounds.

8 These and many other jobs done in the liver produce heat, which the blood carries around your body to keep it warm.

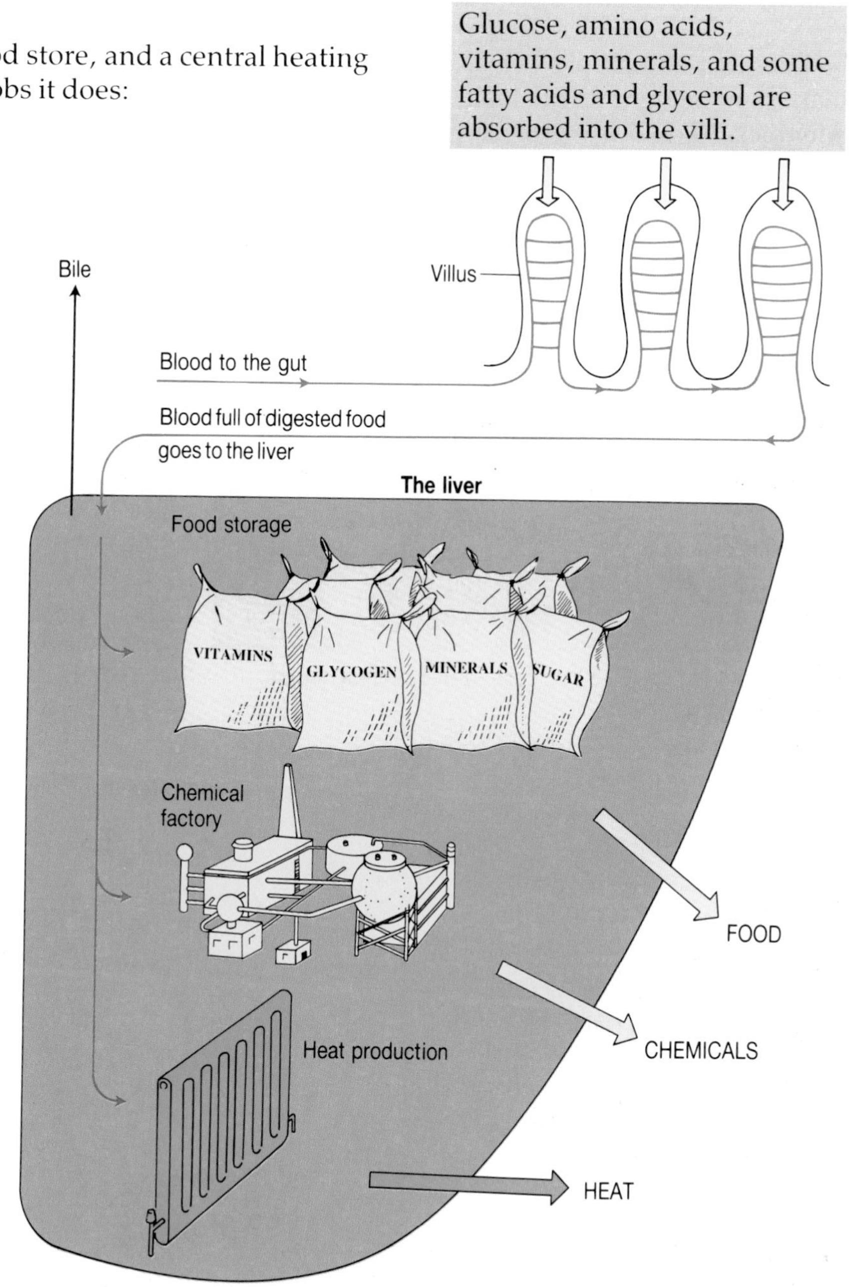

Glucose, amino acids, vitamins, minerals, and some fatty acids and glycerol are absorbed into the villi.

Questions

1. Digested food is absorbed inside your body. What does that mean?
2. Think of a reason why food has to be digested before it can be absorbed.
3. Where in your body are villi? What are they for?
4. Where in your body is your liver?
5. Name three things stored in your liver.
6. How does your liver help give you a steady supply of glucose?
7. Explain how:
   a) your liver helps digestion
   b) your liver helps cuts to stop bleeding
   c) your liver keeps you warm.

6·8 Biotechnology

Both of the photographs below show chemical factories. But the shapes on the right do not look like factories. In fact they are **microbes**. Microbes are tiny living things that can only be seen using a microscope. The microbes in the photograph are **bacteria**, and scientists use them to make chemicals, food, fuels, medicines, and even plastics. Just like a factory.

This chemical factory covers several thousand square metres.

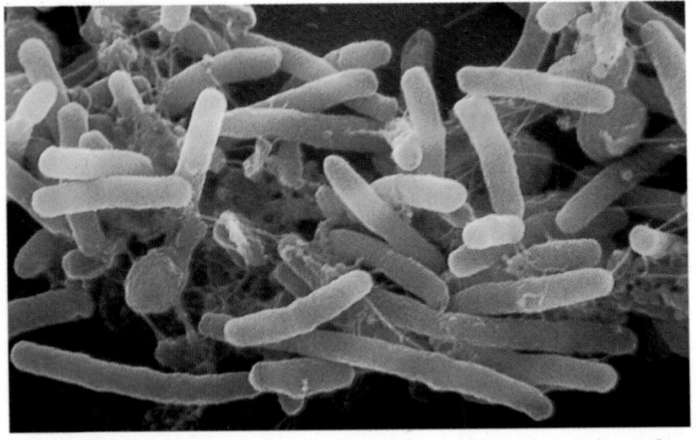

. . . while these are so small that 200 placed end-to-end would fit across this full stop.

Biotechnology

Biotechnology means using microbes and other living things to do jobs for us, and make things we want. For centuries we have used bacteria to make cheese and yoghurt from milk, and other microbes called **yeast** to make bread, beer, and wine.

This diagram shows the many ways that microbes can be used. Because they can produce foodstuff, medicines and fuel, scientists hope that microbes will help to solve the world's food, health and energy problems.

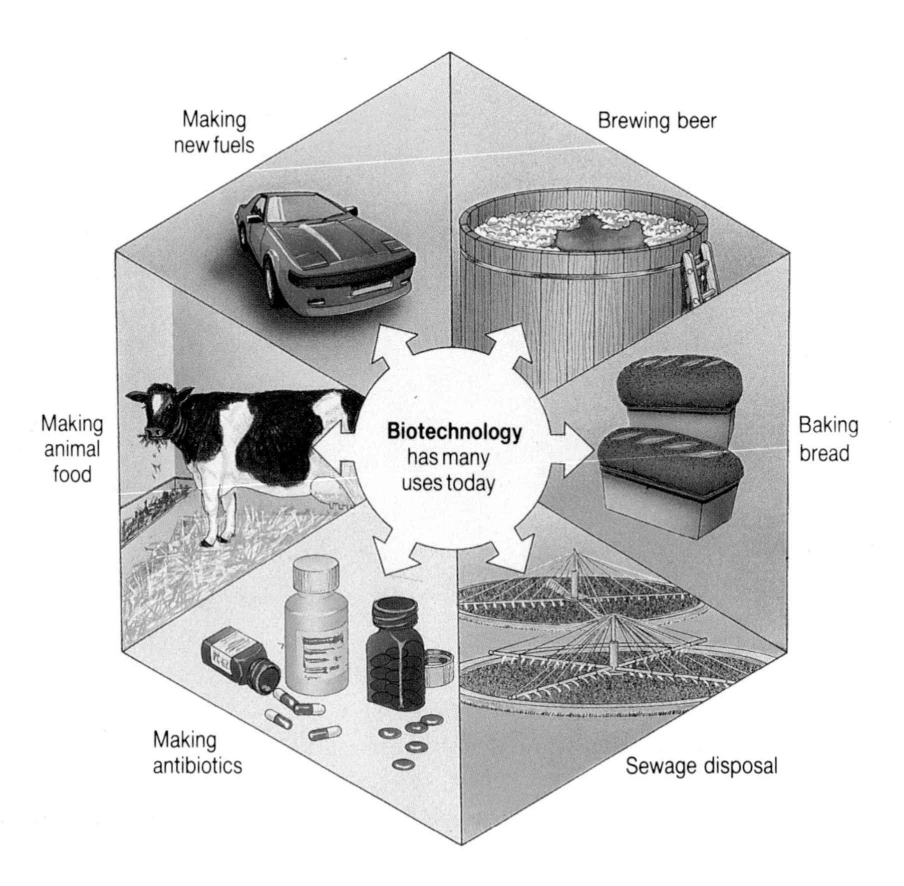

Modern biotechnology

These days biotechnology helps us in many ways.

Medicine. A very powerful medicine called **penicillin** was discovered in 1928. Penicillin is produced by a microbe called a fungus. It is an **antibiotic** which means it kills germs inside the human body.

Food from crude oil. There are microbes that live on cheap chemicals found in crude oil. They form a thick broth which can be dried and compressed to make protein food for animals.

Food from water. Microbes called algae grow in water containing dissolved minerals. They can be used to produce unlimited amounts of proteins, vitamins, and other useful substances for a hungry world.

Gas and fertilizers from sewage. Modern sewage disposal plants use microbes to change sewage into fertilizer. Microbes are used to digest sewage, and animal waste from farms, to produce methane gas. This can be used as fuel in factories and homes.

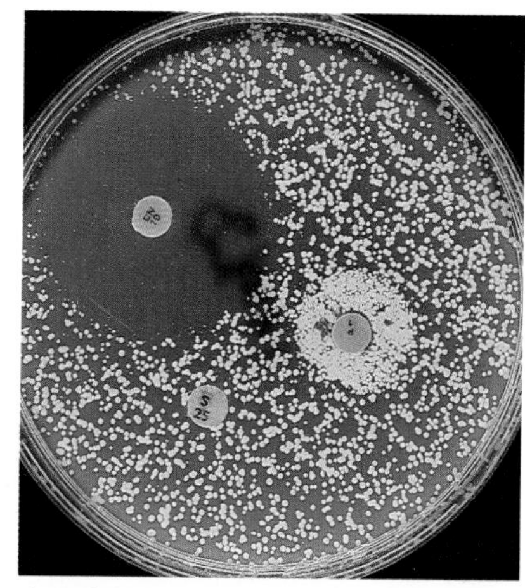

Antibiotics are tested on plates of germs. If an antibiotic is effective it kills a wide radius of germs.

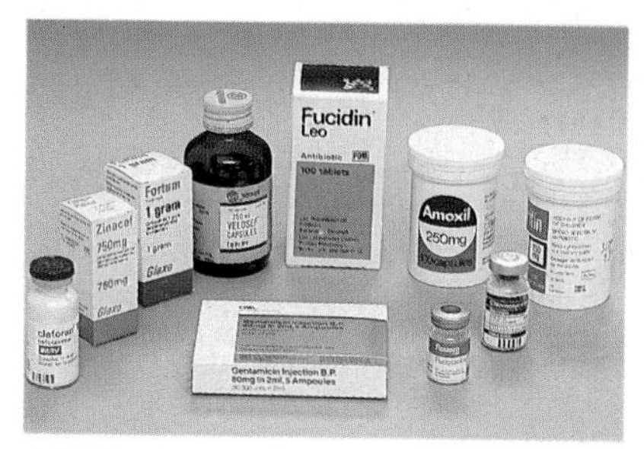

All these antibiotics have been made by microbes.

Genetic engineering

Scientists can make microbes and other organisms produce useful things by changing their genes. This is called **genetic engineering**.

Genes can be transferred from human cells to the plasmids (or gene circles) of bacteria. The genes then work as if they were still in the human cells.

If a gene which controls the production of a human hormone is put into a bacterium, the bacterium becomes a tiny hormone factory.

In this way, bacteria have been used to produce human growth hormones for children who do not grow properly, human insulin for diabetics, and vaccines and vitamins.

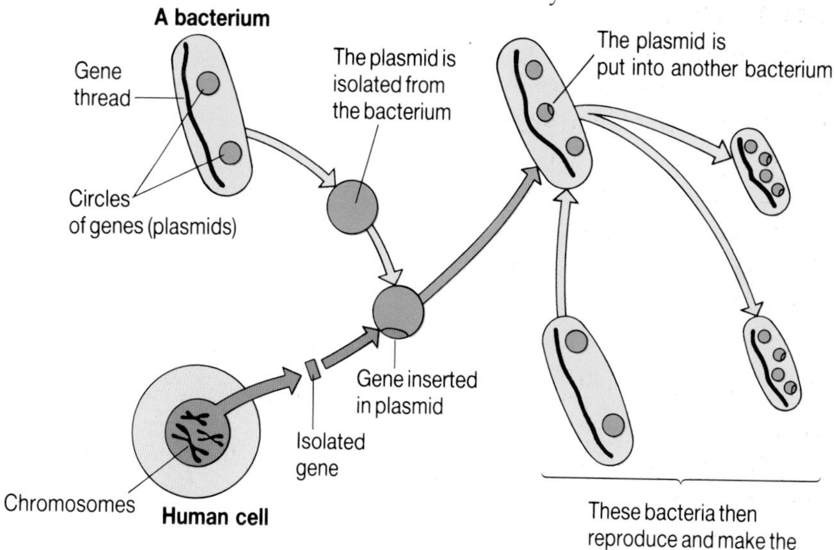

Questions

1 What are microbes? Give two examples.

2 What does biotechnology mean?

3 What are the oldest examples of biotechnology?

4 How could biotechnology help solve the world's food shortage problems?

5 **a)** What is genetic engineering?
b) How can we benefit from it?

Questions

1 Copy the following sentences into your exercise book. Then fill in the gaps using words from the list: **carbohydrates**, **proteins**, **fats and oils**, **vitamins and minerals**, **dietary fibre**.

a) . . . . are body-building foods.
b) Foods rich in . . . . are stored in your body to provide energy in the future.
c) You need . . . . to avoid constipation.
d) Food rich in . . . . provides energy quickly.
e) You need about twenty different . . . . to stay healthy.
f) Starch and sugar are examples of . . . .
g) Fresh fruit contains plenty of . . . .
h) Wholemeal bread contains plenty of . . . .
i) Eggs and fish are rich in . . . .
j) Butter is rich in . . . .

2 **a)** Write down three headings: *Rich in proteins, Rich in carbohydrates, Rich in fats.*
b) Now sort this list of food into three groups under these headings:
jam, lean ham, peanuts, toffee, lard, cake, chicken, egg, butter, potato, fish.
c) Which of the foods in the list:
rot your teeth?
build up cholesterol in your arteries?
make you fat if you eat too much?

3 Copy these sentences, then fill in the gaps.
a) . . . . . and Vitamin . . . . . are needed for strong bones and teeth.
b) Vitamin . . . . . is needed to heal wounds quickly.
c) . . . . . is needed for making red blood cells.
d) Vitamin . . . . . is made in your skin when you sunbathe.
e) Lack of Vitamin . . . . . causes scurvy.
f) Lack of Vitamin . . . . . causes rickets.
g) Lack of . . . . . causes anaemia.

4 Instant mashed potato is an example of a processed food. To make it, potatoes are put through special processes in a factory.
a) What good things may be taken out of food when it is processed?
b) Name three types of food additives which may be added to food during processing.
c) Why are these chemicals added to food?
d) Think of four reasons why people buy processed foods.
e) Think of four reasons why it is better to buy fresh foods than processed foods.
f) What is the difference between:
wholemeal bread and white bread?
juice from a crushed orange and an orange-flavoured drink?
free-range eggs and battery eggs?
garden peas and tinned peas?
roast jacket potato and instant potato?

5 Below is a diagram of the digestive system.
a) Name the parts labelled A to J.
b) Match the labels on the diagram with the descriptions below:
where digested food passes into the blood stream;
where food is mixed with gastric juice;
where faeces leave the body;
a gland which produces enzymes that digest protein, starch, and fat;
glands which produce saliva;
where water is absorbed from the undigested remains of food;
pumps bile into the small intestine;
a tube from the mouth to the stomach;
stores glycogen, vitamins, and minerals.

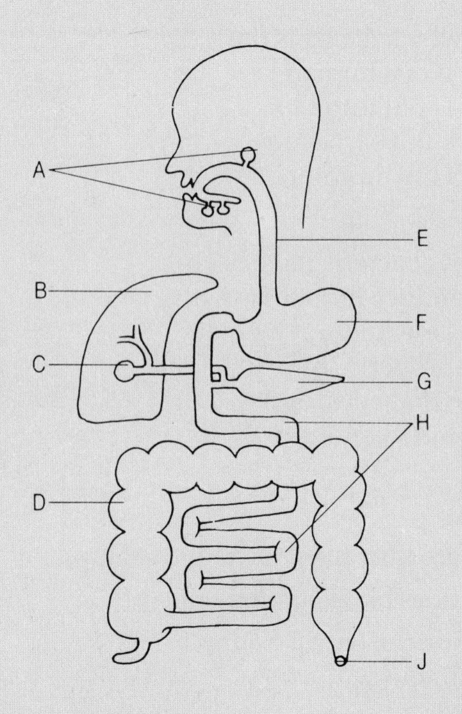

1 Food tests

These tests will tell you which foods contain glucose, starch, fats, oils, and proteins.
You will need these foods:
beans, carrots, and grapes which have been ground up or crushed in a little water; pieces of potato and bread; ground peanuts, butter and suet; milk.

Test for glucose

a) Put equal amounts of Benedict's solution and glucose solution (about 2 cm^3 of each) into a test tube. Lower the tube into a beaker of boiling water.
What do you notice?

b) Repeat the test using the crushed carrots and grapes instead of glucose solution. What can you tell from the results?

Test for starch

a) Add a few drops of iodine solution to some starch solution. What colour appears?

d) Now add iodine solution to the bread, the crushed beans and the cut surface of a potato. What does the test tell you about these foods?

Test for fats and oils

a) Add a little ethanol to a small amount of cooking oil. Shake the mixture and allow it to settle. Then pour off the ethanol into an equal volume of water. What do you notice?

b) Repeat the test, using ground peanuts, butter, and suet instead of cooking oil. What can you say about these foods?

Test for proteins

Add 2 cm^3 of sodium hydroxide solution to milk and mix. Then add a few drops of weak copper sulphate solution. What do you notice? This test shows that milk contains soluble proteins.

2 To show how food moves along the alimentary canal

a) Obtain a length of inner tube (from a bike), and a round stone about the same diameter as the tube.

b) Wash out the tube with water and a little washing-up liquid.

c) Push the stone into the tube.

d) Now curl your thumb and forefinger round the tube, just behind the stone. Use them to squeeze the tube and move the stone along.

e) What does the tube represent?

f) What does the stone represent?

g) What do your thumb and finger represent, when they squeeze the tube?

3 To demonstrate digestion

In your stomach, an enzyme called **pepsin** digests protein. Pepsin needs weak acid to make it work. This test shows how pepsin digests the protein in egg white.

a) Boil an egg for five minutes. Cut it open and remove a large lump of solid egg white.

b) Put the lump of egg white into a test tube and cover it with pepsin solution. Add one drop of dilute hydrochloric acid and make a note of the time. Stand the tube in a beaker of warm water, shaking it from time to time. The egg white disappears as the pepsin digests it. How long does it take to disappear completely?

c) Now separate the white from a raw egg and add it to 500 cm^3 of water in a beaker. Heat the mixture, while stirring, until it turns white.

d) Mix 2 cm^3 of pepsin solution with 2 cm^3 of the white liquid, in a test tube. Add a drop of dilute hydrochloric acid. Put the tube in a beaker of warm water for a few minutes, shaking it from time to time. The white colour will disappear as the pepsin digests the protein in the egg white.

e) This experiment shows that it is important to chew food properly. Explain why.

f) How could you prove that pepsin needs acid to work?

g) Pepsin has digested solid egg white and changed it into a liquid. Why must this happen before your body can use the egg white as food?

7·1 Touch, taste, and smell

You can touch, taste, smell, hear, balance, and see because of your sense organs. In humans the sense organs are:

1 The **skin**, which is sensitive to pressure, heat, and cold and gives you your sense of touch.
2 The **tongue**, which is sensitive to chemicals in food and drink, and gives you your sense of taste.
3 The **nose**, which is sensitive to chemicals in the air, and gives you your sense of smell.
4 The **ears**, which are sensitive to sounds and movement, and give you your senses of hearing and balance.
5 The **eyes**, which are sensitive to light, and give you your sense of sight.

When a sense organ detects a **stimulus** such as sound or light, it sends messages along nerves to the brain. The brain then gives you feelings or **sensations** such as hearing or sight.

A cat has a special set of sense organs – its whiskers. They are the same width as the cat's body. It uses them to judge whether it can squeeze through a gap.

The skin

Your sense of touch is produced by the ends of nerve cells in your skin. These are called **nerve endings**.

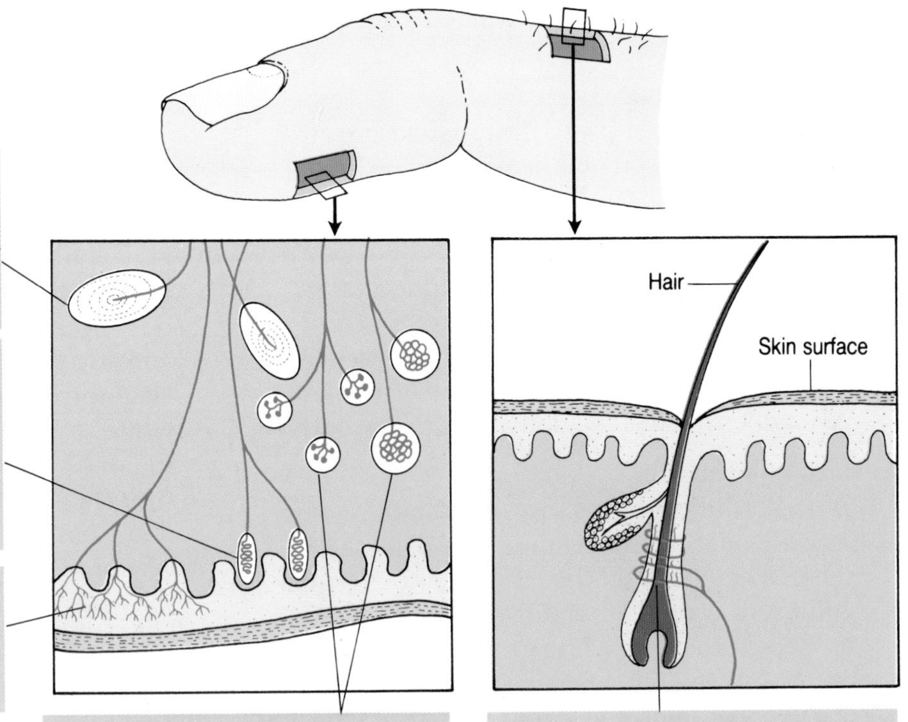

These nerve endings deep inside the skin are sensitive to heavy pressure. They warn you about pressure which could bruise you.

These nerve endings near the skin surface are sensitive to light pressure. They tell you about the texture of an object, for example whether it is rough or smooth.

These nerve endings very close to the skin surface make you feel pain if, for example, your skin is cut or burnt.

These nerve endings are sensitive to heat and cold. They detect changes in temperature. For example, when the weather changes, or when you touch cold or hot objects.

These nerve endings wrapped around the base of a hair detect if the hair is moved or pulled.

The tongue

Your tongue has little bumps on it. These bumps contain tiny sense organs called **taste buds**. Taste buds are sensitive to chemicals in food. These chemicals must dissolve in saliva before you can taste them. This is why dry food has no taste until you chew it to mix it with saliva.

Your sense of taste is useful.

1 It stimulates your stomach to produce gastric juice for digestion.

2 Many poisons and bad foods have a nasty taste. So you can spit them out before they harm you.

There are different taste buds for tasting bitter, sour, salty, and sweet foods.

A food can stimulate more than one kind of taste bud at the same time.

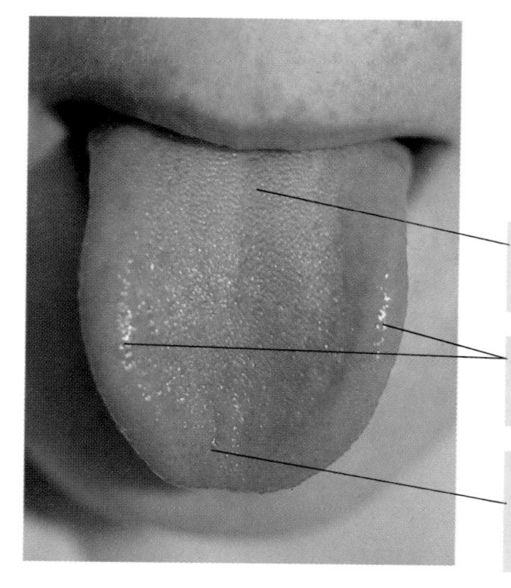

The taste buds for **bitter tastes** are at the back of your tongue.

The taste buds for **sour tastes** are at the sides of your tongue.

The taste buds for **salty and sweet** tastes are at the front of your tongue.

The nose

When a bad cold blocks your nose, food seems to lose its flavour. This is because many flavours are really smells!

Smells are chemicals in the air. The chemicals dissolve in moisture on the lining of your nose. This stimulates nerve endings in your nose to send messages to your brain which produces the sensation of smell.

Humans can detect about 3000 different smells. Smells help animals to hunt food and find their way. Smells can also warn of certain dangers.

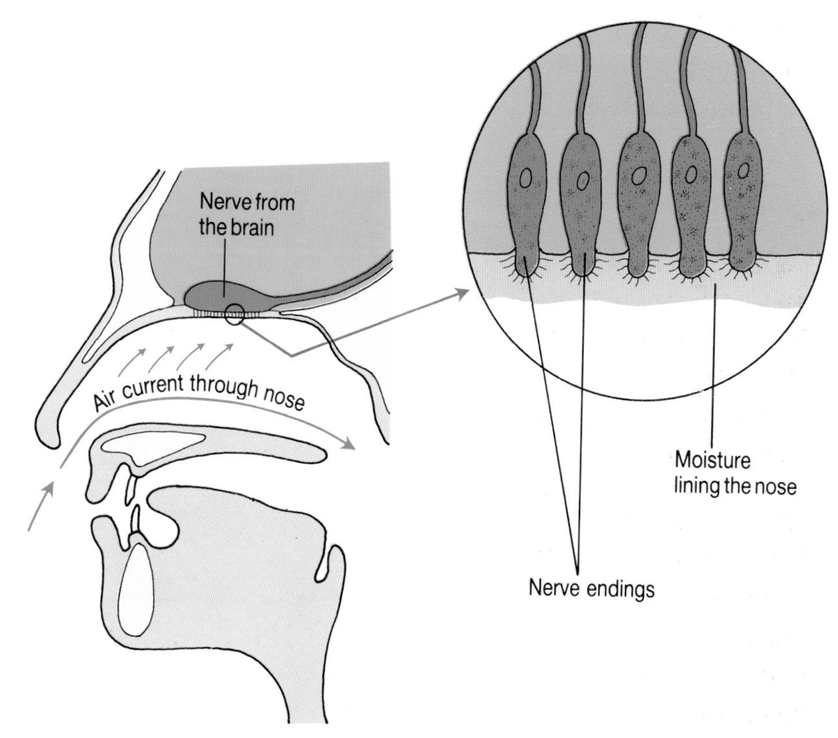

Questions

1 Name the five sense organs in humans.

2 If you gently stroke the hairs on your arm you feel a tingling sensation. Try to explain why.

3 Why do foods seem to have less flavour when you have a cold in the nose?

4 How do taste and smell protect you?

7·2 The eye

How you see things

1 Light goes from an object to your eye.

2 Light is bent as it passes through your eye.

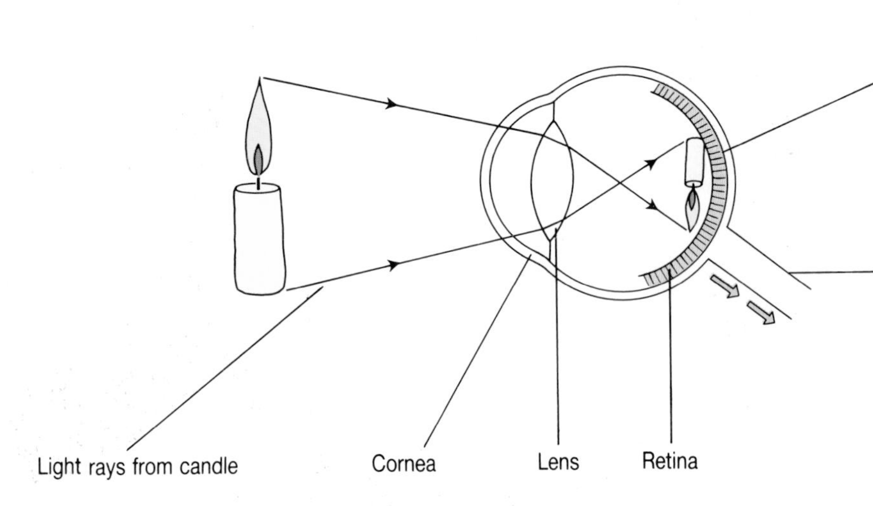

3 An upside-down picture of the object is focused on a layer called the **retina** at the back of your eye. The retina is made of cells sensitive to light.

4 The retina sends messages along the **optic nerve** to your brain. Your brain allows you to see a picture of the object which is the right way up.

How eyes are protected

Your eyes are set in holes called **orbits** in your skull. So all except the front of each eye is protected by bone.

There are **tear glands** behind the top eyelid. They make tears that wash your eye clean when you blink. Tears are produced faster if dust or smoke get into your eyes.

Eyelashes form a net in front of the eye which protects it from dust.

The white of the eye is a tough protective layer called the **sclerotic.**

The **iris** is the coloured part of the eye. It is a ring of muscle with a hole called the pupil in the middle. It protects the eyes from bright light.

The **pupil** lets light into the eye. If the light is too bright the iris muscle makes the pupil smaller. In dim light the iris muscle make the pupil bigger.

The parts of the eye

The **conjunctiva** is a thin clear skin which covers the front of the eye.

The **cornea** is a clear window in the sclerotic in front of the iris. It lets light into the eye.

The **iris** controls the amount of light entering the eye.

The front part of the eye is filled with a watery liquid called **aqueous humour**.

The **lens** helps focus a picture on the retina. The lens is clear and can change shape.

The **suspensory ligaments** hold the lens in place.

The **sclerotic layer** is the tough, white protective layer of the eye.

The back of the eye is filled with a jelly called **vitreous humour**.

The **choroid** is a black layer that stops light being reflected round the inside of the eye.

Ciliary muscles change the shape of the lens during focusing.

The **yellow spot** is the most sensitive part of the retina. It lets you see colour.

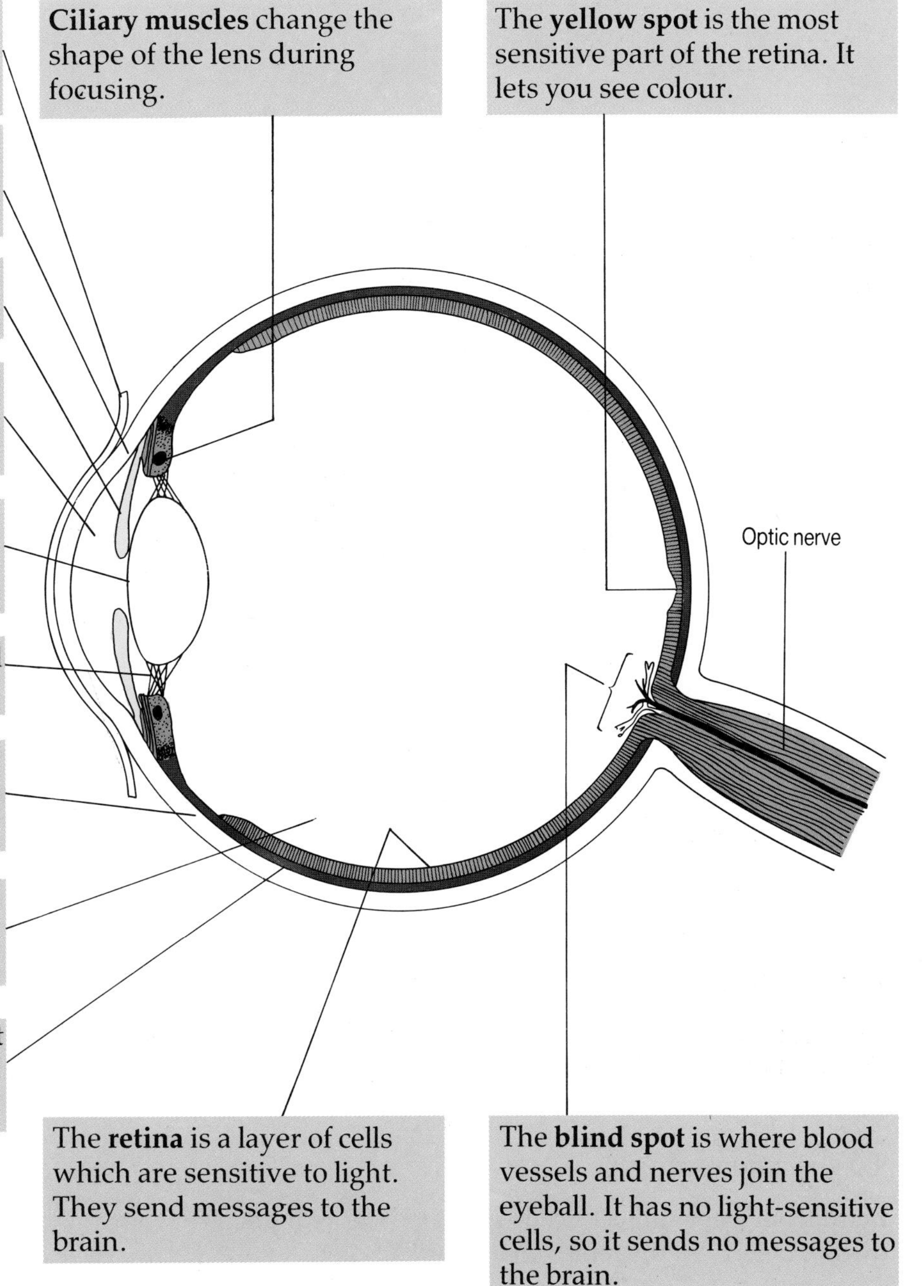

The **retina** is a layer of cells which are sensitive to light. They send messages to the brain.

The **blind spot** is where blood vessels and nerves join the eyeball. It has no light-sensitive cells, so it sends no messages to the brain.

Questions

1 Give the scientific name for each of these:

- **a)** carries messages from an eye to the brain.
- **b)** muscles which change the shape of a lens.
- **c)** A layer of light-sensitive cells.
- **d)** It controls the amount of light entering the eye.
- **e)** It prevents light being reflected around the eye.
- **f)** They make liquid which washes the eyes.
- **g)** A hole in the middle of the iris.
- **h)** A clear window at the front of the eye.
- **i)** changes shape to focus a picture on the retina.

2 If you walk from a dark room into sunlight and back again how would your pupils alter in size? Why does this happen?

7·3 Vision

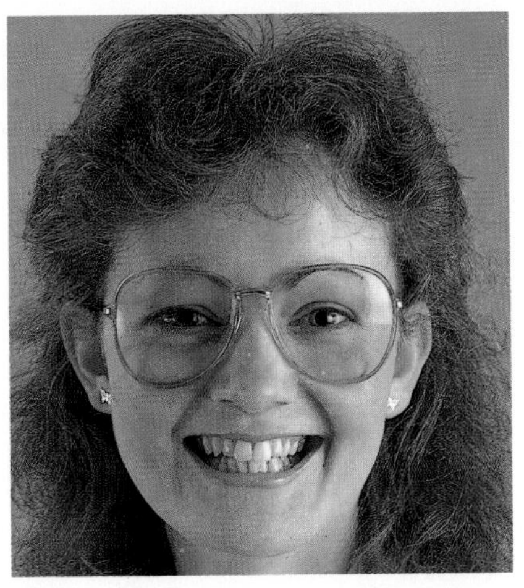

If the lenses in your eyes don't focus properly, wearing glasses will help

The light that goes into your eye has to be bent or **focused** onto the retina to let you see clearly.

Most of the bending of light takes place as it passes through the **cornea** and **aqueous humour**. The **lens** bends it a little more to make a perfectly clear picture on the retina. The **ciliary muscles** change the shape of the lens to bend light. A fat lens bends light more than a thin lens.

To see a distant object

Light from a distant object needs to be bent very little. So the lens is stretched to make it thin.

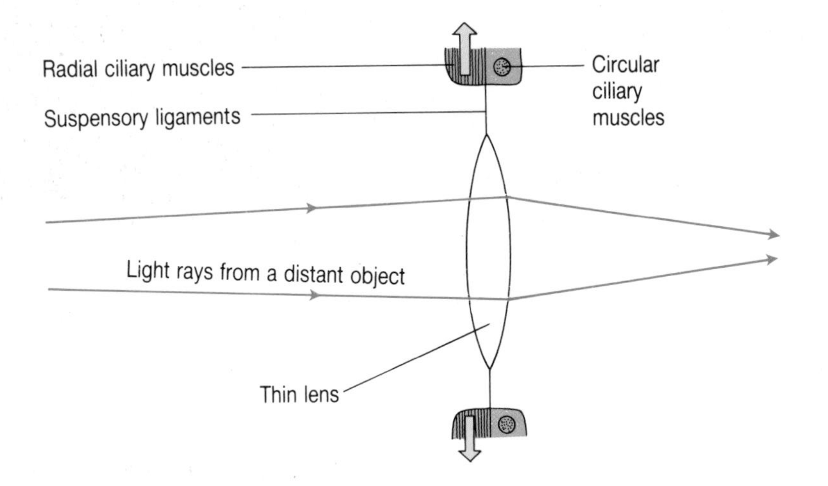

1 The **radial ciliary muscles** contract and pull against the edges of the lens.

2 This stretches the lens making it thin and flat. So it bends light just a little, to make a clear picture on the retina.

To see a near object

Light from a near object needs to be bent more than light from a distant object. So the lens is made much fatter.

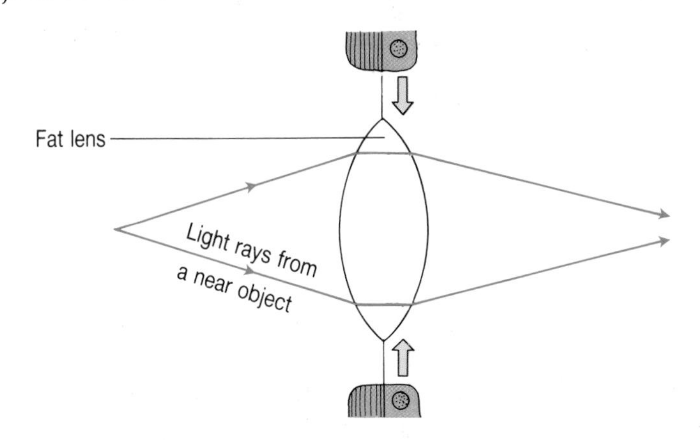

1 The ciliary muscles are in a circle round the eyeball. They contract and squeeze the eyeball inwards.

2 The lens becomes much fatter in shape. Now it can bend light enough to make a clear picture on the retina.

Three dimensional vision

Each of your eyes gets a slightly different view of an object.

Your brain puts these two views together, so that you see the object as three-dimensional rather than flat.

Three-dimensional vision helps you judge how far away an object is.

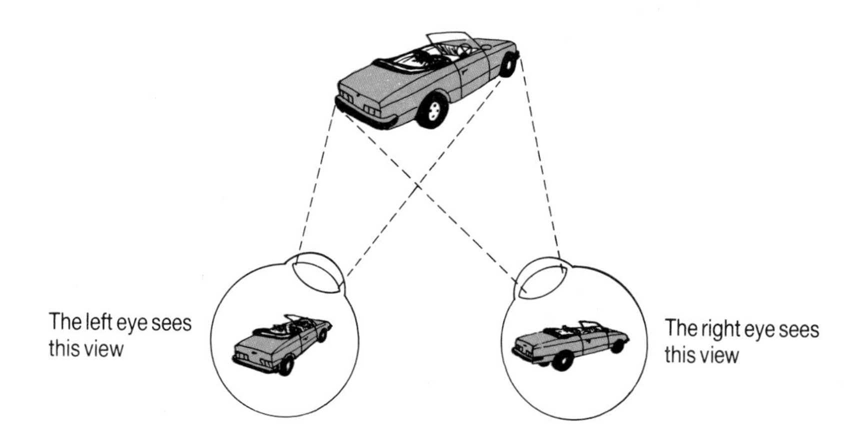

Two eyes

Rabbits, chickens, fish, and many other animals have eyes which look sideways, and not forwards like yours. Each eye sees a different view. They can even see what is happening behind them. This is useful if other animals hunt you for food!

Rabbits can see what's going on at each side...

...but owls look straight ahead.

More about the retina

There are two kinds of light-sensitive cells in the retina. They are called **rods** and **cones**.

Cones only work in bright light, but give a very clear picture and are sensitive to colour. The **yellow spot** in the middle of the retina is made entirely of cones.

If you want to see something very clearly you look straight at it, so that its picture falls on the yellow spot.

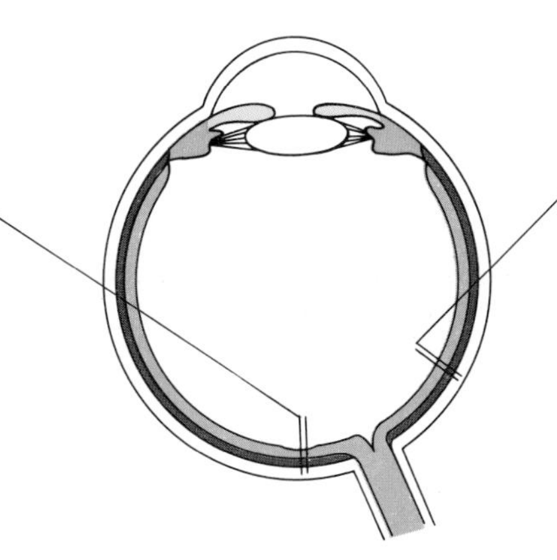

The rest of the retina is mostly rods, with a few cones. Rods do not give as clear a picture as cones and are not sensitive to colour. But rods work in dim light.

This explains why you don't see colours clearly in dim light.

Questions

1 Which parts of an eye bend light?

2 What shape are your eye lenses when you look at your hand? What shape are they when you look at a tree in the distance?

3 **a)** Where in the retina are rods and cones found?

b) Which detect colour, and which work in dim light?

7·5 The nervous system

Your nervous system is like a manager inside your body. Its job is to **control** and **co-ordinate** the parts of your body so that they work together, doing their jobs at the right time.

Your nervous system co-ordinates muscles so that you can do things which need thought, like cycling, dancing, or reading.

It also co-ordinates things which you don't need to think about, like heart beat and breathing.

The nervous system consists of the **brain**, the **spinal cord**, and millions of **nerves**. Together the brain and spinal cord are called the **central nervous system**.

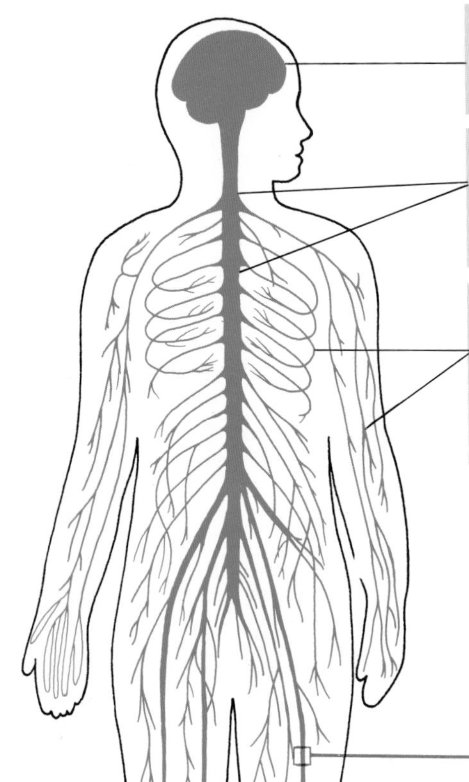

The **brain** is at the top of the nervous system. It is protected by the skull.

The **spinal cord** is a very thick nerve. It hangs from the brain down through the hollow middle of the backbone.

Millions of **nerves** branch from the central nervous system. They carry messages called **nerve impulses** around the body.

A nerve is a bundle of **nerve fibres**.

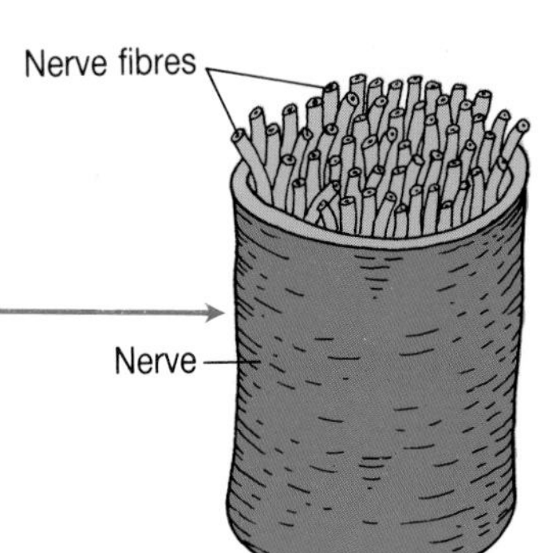

Nerve cells

The nervous system is made up of **nerve cells**. Most cells are small and rounded. But nerve cells are not. They are stretched out into long thin **nerve fibres** that can be over a metre long. Nerve impulses travel along nerve fibres in *only one direction*.

Sensory nerve cells carry impulses from sense organs into the central nervous system.

Motor nerve cells carry impulses from the central nervous system to muscles and glands.

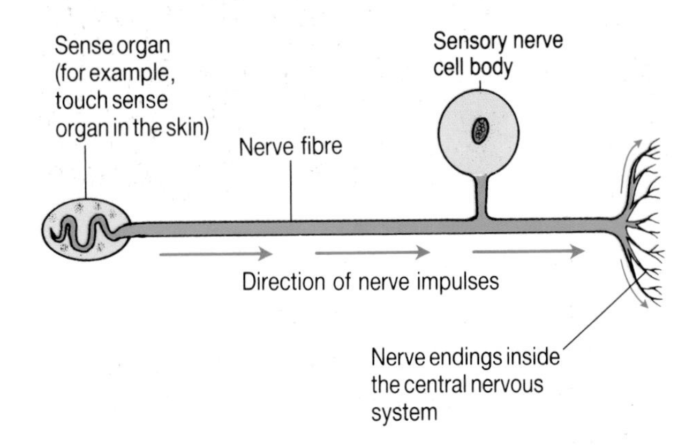

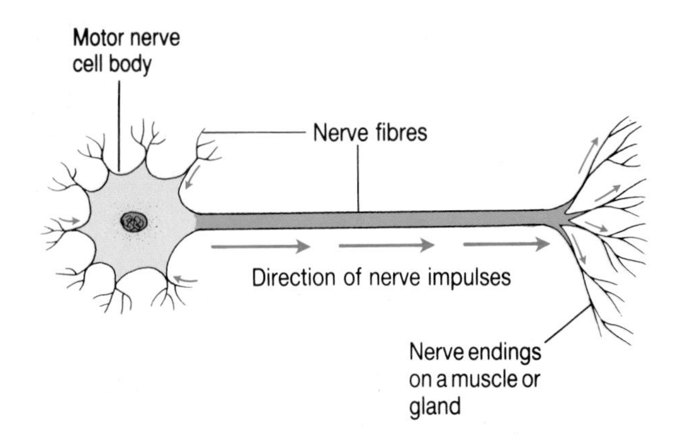

How the nervous system works

If you sat on a drawing pin you would jump up yelling with pain. This is an example of a **stimulus** and a **response**. The stimulus is pain. The response is jumping and yelling. Your nervous system controls the response.

This diagram of the nervous system shows how it works.

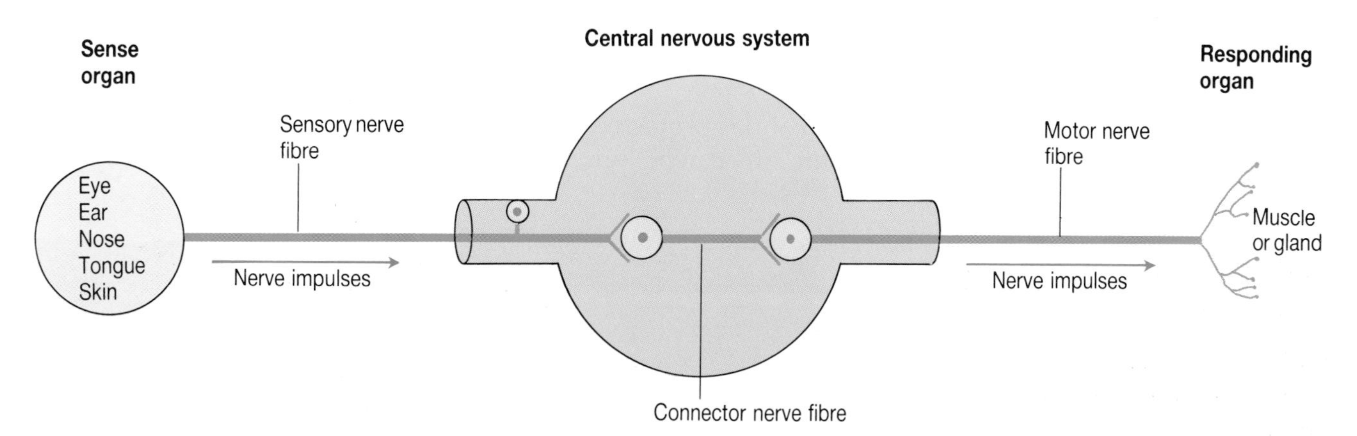

1 A stimulated sense organ sends nerve impulses along **sensory nerve fibres** to the central nervous system.

2 The central nervous system works out the best response to the stimulus. Then it sends impulses to the muscles and glands which will carry out the response.

3 The impulses travel to the muscles and glands along **motor nerve fibres**.

Reflex and voluntary actions

Reflex actions are actions you do without thinking, to protect yourself: coughing clears your windpipe, and shivering keeps you warm. The pupils get smaller in bright light, to protect the retina and larger in dull light to help you see.

Voluntary actions are actions which need thought, like speaking to a friend or writing a letter.

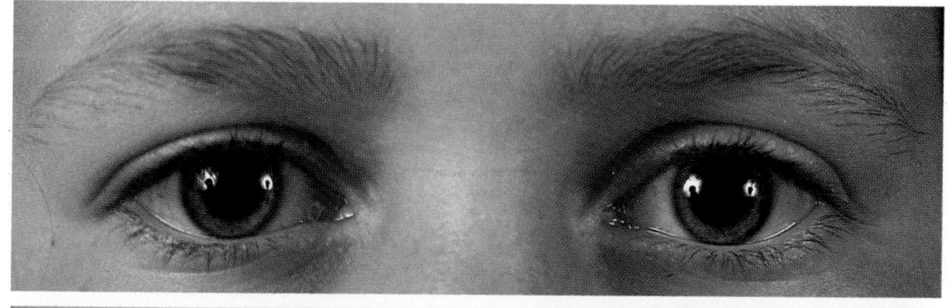

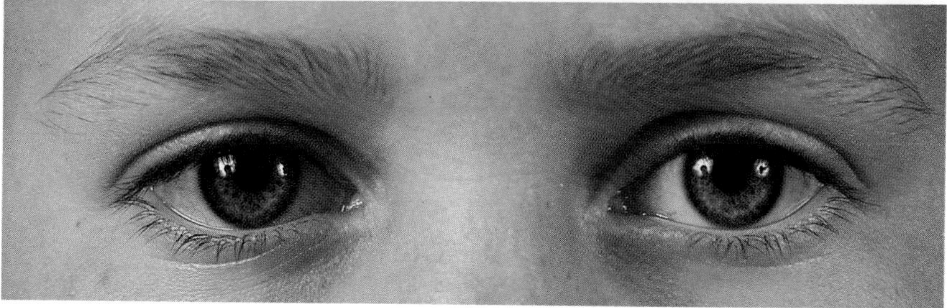

Questions

1 Name the parts of the nervous system.
2 Name two kinds of nerve cell.
3 How are nerve cells different from other cells?
4 What is a nerve impulse?
5 What is a reflex action?
6 What is a voluntary action?
7 Sort these into reflex and voluntary actions: coughing, reading, sneezing, sweating, writing.

7·5 The brain

There are about 10 000 million nerve cells in your brain. Each cell is linked with thousands more. This linking allows your brain to do many different jobs all at the same time.

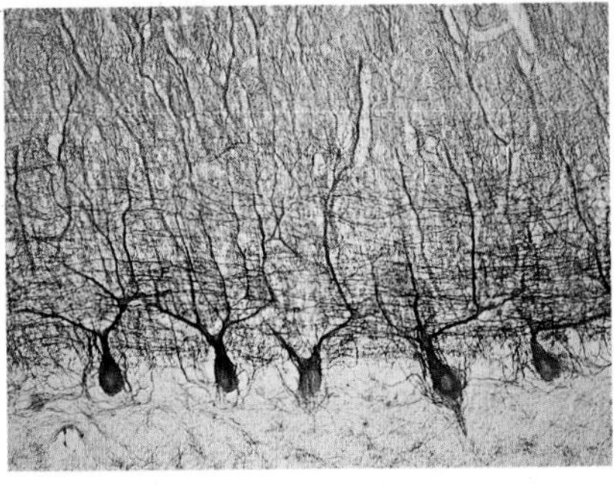

Nerve cells from the cerebellum of the brain. Their job is to control muscles and balance.

The parts of the brain

Your brain is at the top of your spinal cord and is protected by your skull. It has three main parts: the cerebrum, the cerebellum, and the medulla.

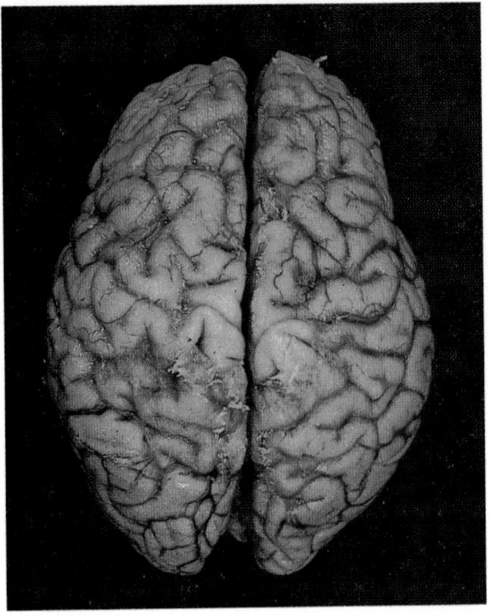

Your brain looks like this from above. The cerebrum is divided into two halves, called the cerebral hemispheres.

The **cerebrum** is the dome-shaped roof of the brain.

Its **sensory areas** (blue) receive impulses from your eyes, ears, tongue, nose, and skin, and give you sensations or feelings.

Its **motor areas** (red) control your muscles during movement.

Its **association areas** (yellow) control memory and thinking.

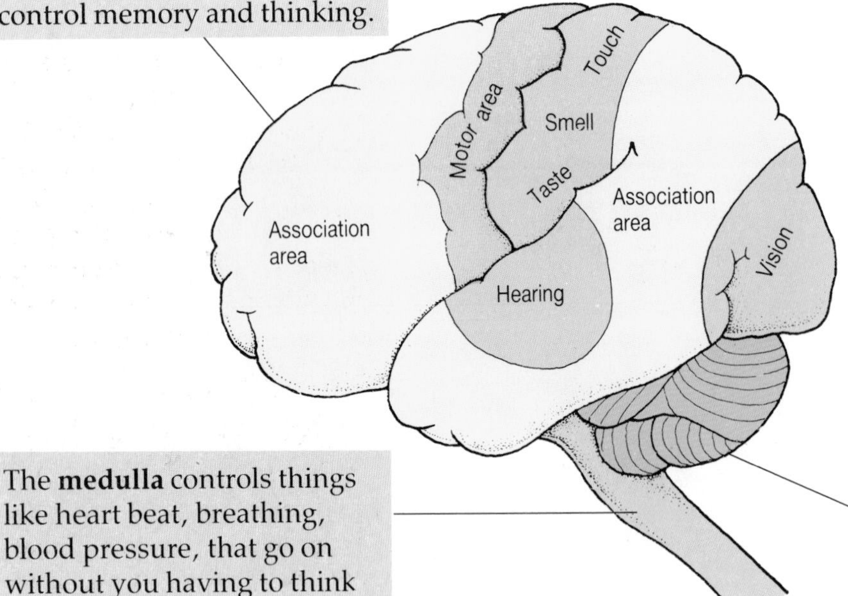

The **medulla** controls things like heart beat, breathing, blood pressure, that go on without you having to think about them.

The **cerebellum** helps control your muscles and balance during walking, running, cycling, and so on.

Question

1 Which part of the brain:
  a) do you use to think?
  b) controls your breathing?
  c) helps you balance while cycling?
  d) gives you the sensation of touch?

7·6 Hormones

Hormones are made in a set of glands called the **endocrine glands**. Like the nervous system, hormones co-ordinate the body.

Endocrine glands release tiny amounts of hormones into the blood. Blood carries them all over the body, but only certain parts called **target organs** respond to them. Responses to hormones may last a few minutes, or may go on for years.

This diagram shows the main endocrine glands, the hormones they produce, and the jobs which the hormones do.

The **thyroid gland** is attached to the windpipe. It makes a hormone called **thyroxine**.

Thyroxine controls the speed of chemical reactions in cells. In children, too little thyroxine leads to slow growth and mental development.

The **adrenal glands** are on top of the kidneys. When you are angry or frightened they make a hormone called **adrenalin**.

Adrenalin prepares your body for action. It speeds up heart beat and breathing, raises blood pressure, and allows more glucose to go into the blood, to give you energy.

In females the **ovaries** make a sex hormone called **oestrogen**. Oestrogen gives girls their female features such as breasts, soft skin, and a feminine voice. It also gets the womb ready for a baby.

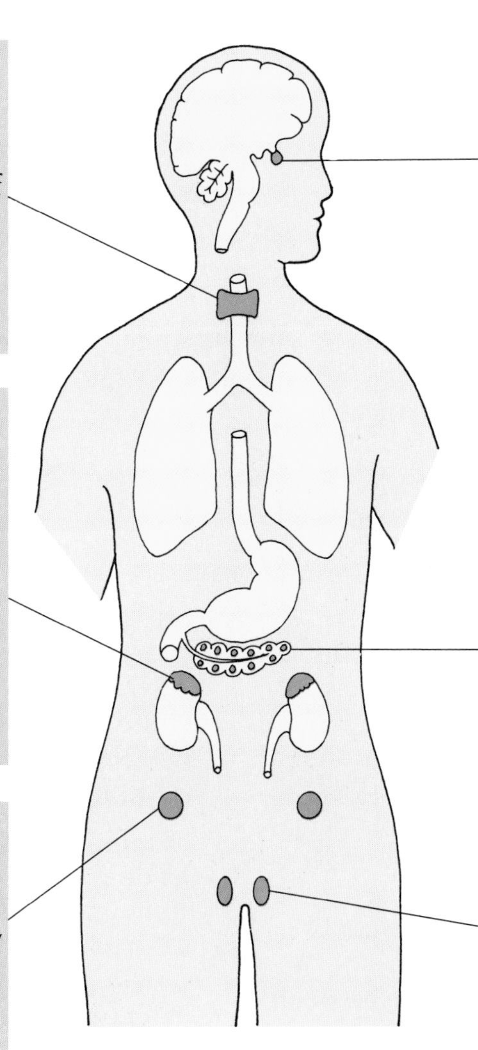

The **pituitary gland** is under the brain. It makes many hormones.

1 One of these hormones controls growth. A person with too little growth hormone becomes a dwarf. A person with too much becomes a giant.

2 In females, it makes hormones which control the release of eggs from ovaries, and the birth of a baby.

The **pancreas** is below the stomach. The pancreas produces a hormone called **insulin**. Insulin controls the amount of glucose in the blood. People with an illness called **diabetes** make too little insulin. So their liver releases harmful amounts of glucose into the blood.

In males the **testes** make a sex hormone called **testosterone**. This gives boys their male features such as deeper voices and more body hair than females.

Question

1 Which hormone:
- **a)** prepares the body for action?
- **b)** controls the amount of glucose in blood?
- **c)** gives boys a deep voice?
- **d)** gives girls soft skin?
- **e)** controls chemical reactions in cells?

Questions

1 Three students were asked to sit in a row with their tongues out.
At the same moment crystals of sugar were placed on the front of the first tongue, on the sides of the next tongue, and at the back of the third one.

a) Which student was the first to taste the sugar? Explain why?
b) What will the results be if the test is repeated using salt? Using lemon juice? Using bitter-tasting instant coffee?
c) To taste food and drink properly, you should spread it all over your tongue. Explain why?
d) How do taste and smell protect us from harm?

2 **a)** What is a reflex action?
b) What reflex action happens:
when dust blows into your eyes?
when a bright light shines in your eyes?
when a very cold wind suddenly blows over you?
when food 'goes the wrong way' and gets into your wind-pipe?
c) Explain how each of these reflex actions protects you from harm.

3 The diagram below shows the inside of an eye.

a) Name the parts labelled A to L.
b) How does the shape of part J change:
when the eye is focused on a near object?
when it is focused on a distant object?
c) Describe the main differences between the parts of the eye labelled C, D, and E.
d) What happens to part I:
in dim light? in bright light?
e) Describe two functions of part B.
f) What passes from the eye along part F?

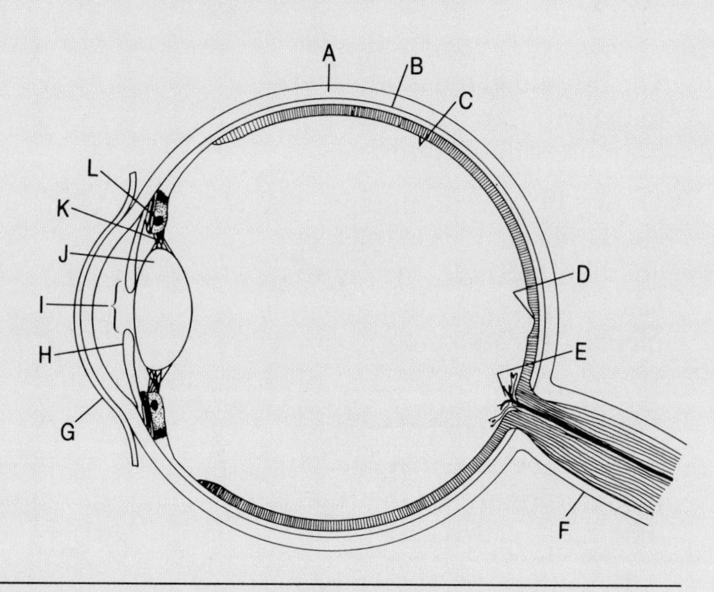

Investigations

1 **Measure your field of vision**
Your 'field of vision' tells how far to each side you can see, when your eyes are looking straight ahead.

a) Fix your eyes on something straight ahead and don't move them.
b) Hold both arms straight out in front of you with thumbs pointing upwards. Now move your arms slowly sideways, until your thumbs are just visible out of the corners of your eyes.
c) Ask a friend to estimate the angle between your outstretched arms. This angle is your field of vision.
d) How would your field of vision differ if your eyes were at the sides of your head, like a chicken?
e) Think of one advantage of having eyes that look forward, like yours.
f) Think of one advantage of having eyes at the side of your head.
g) Name two more animals, other than humans which have eyes that face forwards.
h) Name two more animals, other than chickens which have eyes that face sideways.

2 **Seeing double**
Why do we need two eyes? This investigation will show you one reason.

a) Hold up a coin with its edge towards you, about 30 cm in front of your eyes.
b) Look at the coin with your left eye closed. Now look at it with your right eye closed. What do you notice?
When you look at the coin with both eyes, your brain puts the two different views together, to give you three dimensional vision.

3 Judging distance

This investigation will show you another advantage of having two eyes.

a) Arrange two pencils on a desk top in the positions A and B shown in this diagram.

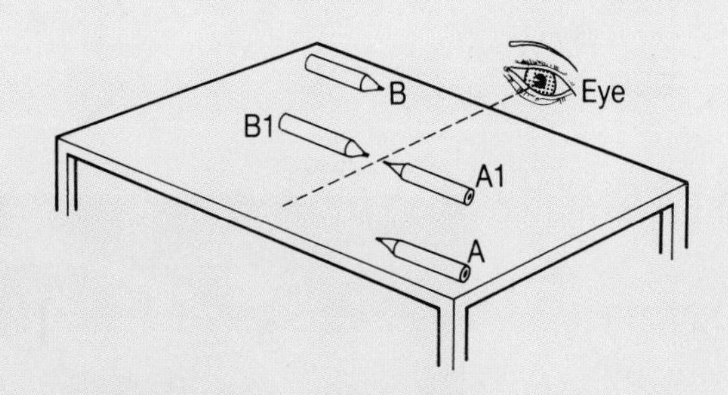

b) Sit so that your eyes are level with the surface of the desk. With one eye closed look across the surface at the pencil point.

c) Using only *one* hand, try to arrange the two pencils so that their points are exactly opposite each other (positions A1 and B1 in the drawing).

d) Return the pencils to positions A and B. Now try the experiment again with both eyes open. Was it easier using one eye?

You should have found the experiment easier with two eyes. This is because both eyes together give you **distance judgement**. They let you judge how far away objects are.

4 Sensing temperature

a) Take three 500 cm^3 beakers. Half-fill one with hot (not boiling) water, the second with ice-cold water, and the third with water at room temperature.

b) Put the fingers of your left hand in the hot water, and the fingers of your right hand in the ice-cold water.

c) After one minute put the fingers of *both* hands in the water that's at room temperature.

d) Does this water feel different to each hand?

e) To which hand does it feel warmer?

How can the same water feel warm *and* cold at the same time? The answer is that your skin is sensitive to *changes* in temperature. Your left hand tells you that the temperature around it has suddenly dropped. But your right hand tells you that the temperature around *it* has risen.

5 How fast do nerve impulses travel?

This investigation will give you an idea of how long it takes for a nerve impulse to travel from one hand, through your nervous system, to your other hand.

a) Stand in a line with the rest of the class, holding hands. Your teacher will stand at one end of the line holding a stop watch. The student at the other end of the line must be ready to bang his or her hand on a desk top.

b) Without warning the teacher will squeeze the hand of the first student, starting the stop watch at the same instant. This student immediately squeezes the next student's hand and so on down the line.

c) When the last student's hand is squeezed, he or she bangs the desk top with the other hand. The teacher instantly stops the stop watch.

d) To find how long it takes for a nerve impulse to travel through one person, do this calculation:

$$\textbf{Time taken for a nerve impulse to travel through one person in the line} = \frac{\textbf{Total time taken}}{\textbf{Number of students}}$$

e) Repeat the experiment several times to get an average result.

A result faster than 0.15 seconds for this experiment is very good.

6 Your ears as direction-finders

a) Blind-fold a student and put her to sit in the middle of your classroom.

b) Ask another student to take off his or her shoes and move very quietly to different parts of the room, making a sharp noise (like a clap) at intervals.

c) The blind-folded student must then try to point in the direction of the claps. Keep a score of her correct and incorrect guesses.

d) You will probably find that it is quite difficult to judge the direction of some claps. Which particular directions do these come from?

e) Look at the shape of the ears and explain why this happens.

f) Repeat the experiment, covering one ear.

g) Repeat the experiment for another group of students and compare your results.

8·1 Two kinds of reproduction

Reproduction is the creation of new living things. There are two ways it can happen: by **asexual reproduction**, or by **sexual reproduction**.

Asexual reproduction

In asexual reproduction there is only one parent. Most tiny simple organisms reproduce in this way.

Tiny creatures like *Amoeba* and *Paramecium* are made up of only one cell. They reproduce asexually in the simplest possible way, by dividing to form two cells.

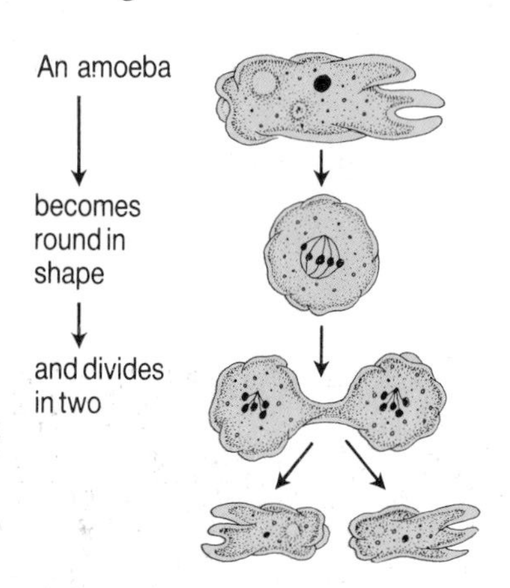

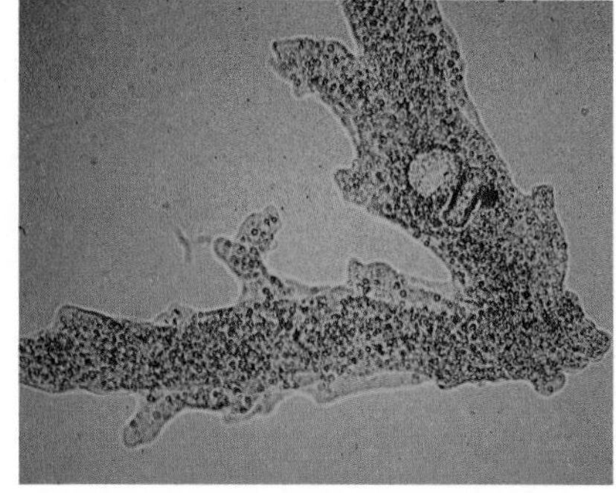

Amoeba takes about an hour to divide. The two parts separate to give two identical organisms.

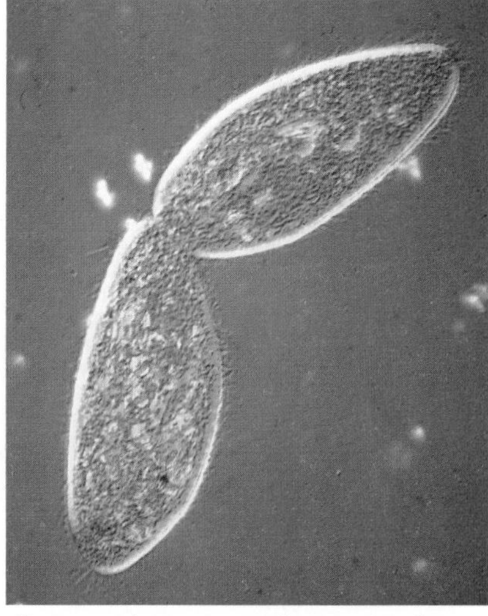

Paramecium reproduces in the same way. Notice the fine hairs or cilia round both parts.

Hydras and sea anemones are made up of many cells. They reproduce asexually by growing buds. A bud starts off as a swelling. Then it develops tentacles, a mouth and a gut, and splits off the parent.

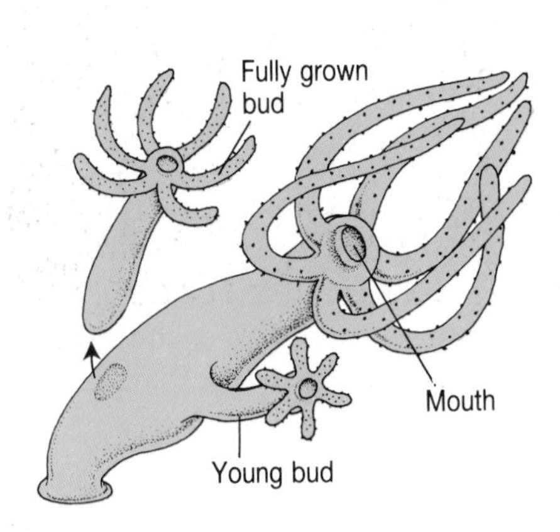

An adult Hydra. The bud on the right has grown its tentacles and is ready to break off.

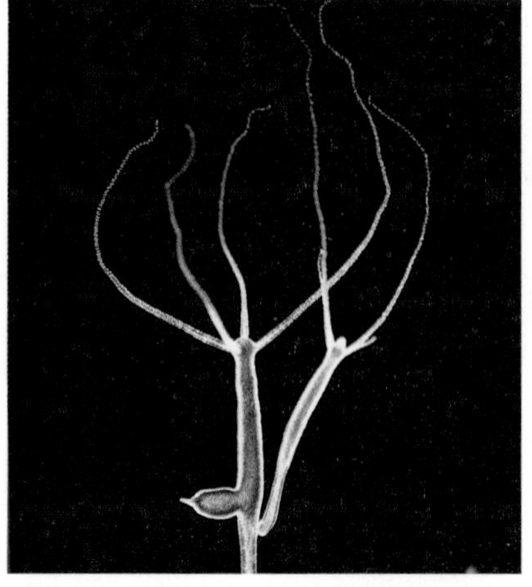

After separating, the new bud floats away, settles down to feed and grows into an adult.

The beadlet sea anemone grows pink buds inside its body, then squirts them out of its mouth.

Sexual reproduction

In sexual reproduction there are two parents. The parents have **sex organs** which produce **sex cells**. In animals the male sex cells are called **sperms**. Sperms are produced by sex organs called **testes**. Female sex cells are called **eggs** or **ova** (one is an ovum). Ova are produced by sex organs called **ovaries**.

Here is an example of sexual reproduction.

Female sticklebacks produce eggs.

Male sticklebacks produce sperms.

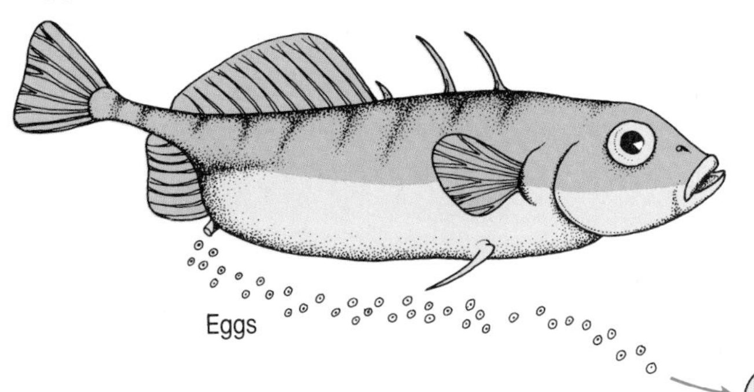

A sperm enters an egg. This is called **fertilization**.

A fertilized egg divides many times to form a ball of cells. The cells develop into a baby stickleback. A partly-formed baby is called an **embryo**.

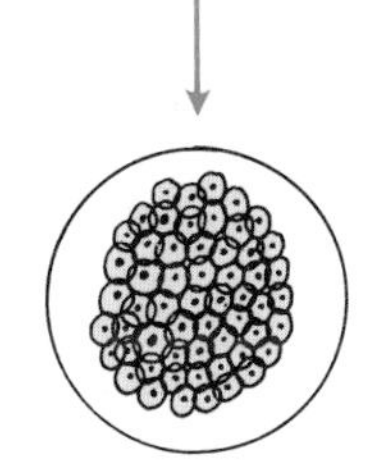

These stickleback embryos are nine days old.

External and internal fertilization

A female stickleback's eggs are fertilized outside her body. This is an example of **external fertilization**.

In insects, reptiles,birds, and mammals (including humans) the male puts his sperm into the female's body. So the eggs are fertilized inside her. This is called **internal fertilization**.

Plants reproduce sexually. You can read about them on page 52.

Questions

1 How is sexual reproduction different from asexual reproduction?

2 **a)** What are male sex cells and female sex cells called?

b) Name the organ which produces male sex cells.

c) Name the organ which produces female sex cells.

8·2 Human reproduction I

Sexual development in boys

Sometime between 11 and 16, a boy goes through these changes:
1 his voice becomes deeper
2 hair starts to grow on his face and body
3 his muscles develop
4 his testes start producing **sperm**.

These changes are caused by a hormone called **testosterone** produced by the boy's testes.

The changes that take place in young people during sexual development are called puberty.

Male sex organs

Male sex organs make male sex cells called **sperms**. They pass them into a woman's body during sexual intercourse.

A **testis** makes sperms and the male hormone testosterone. A man has two testes.

The **penis** passes sperms from the man's body into the woman's body during sexual intercourse.

These spaces are called **erectile tissue**. Before intercourse they fill with blood, making the penis stiff and erect.

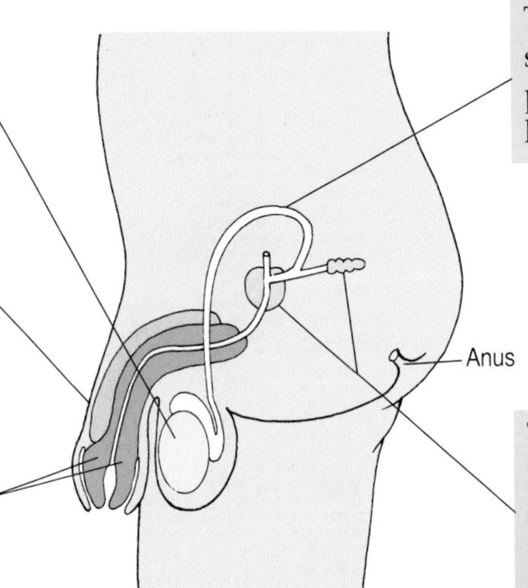

These **sperm tubes** carry sperms from the testes to the penis. Sperms are carried in a liquid called **semen**.

These **glands** make the liquid part of semen. It contains chemicals which cause the sperms to swim after they enter the woman's body.

Sexual development in girls

Sometime between 8 and 15, a girl goes through these changes:
1 her breasts get bigger
2 her hips get more rounded
3 hair starts to grow on parts of her body
4 her ovaries start releasing eggs (ova).

These changes are caused by a hormone called **oestrogen** produced by the girl's ovaries.

Girls develop sexually earlier than boys do. But of course it varies from person to person.

Female sex organs

Female sex organs make ova, and protect and feed an ovum if it is fertilized and develops into a baby.

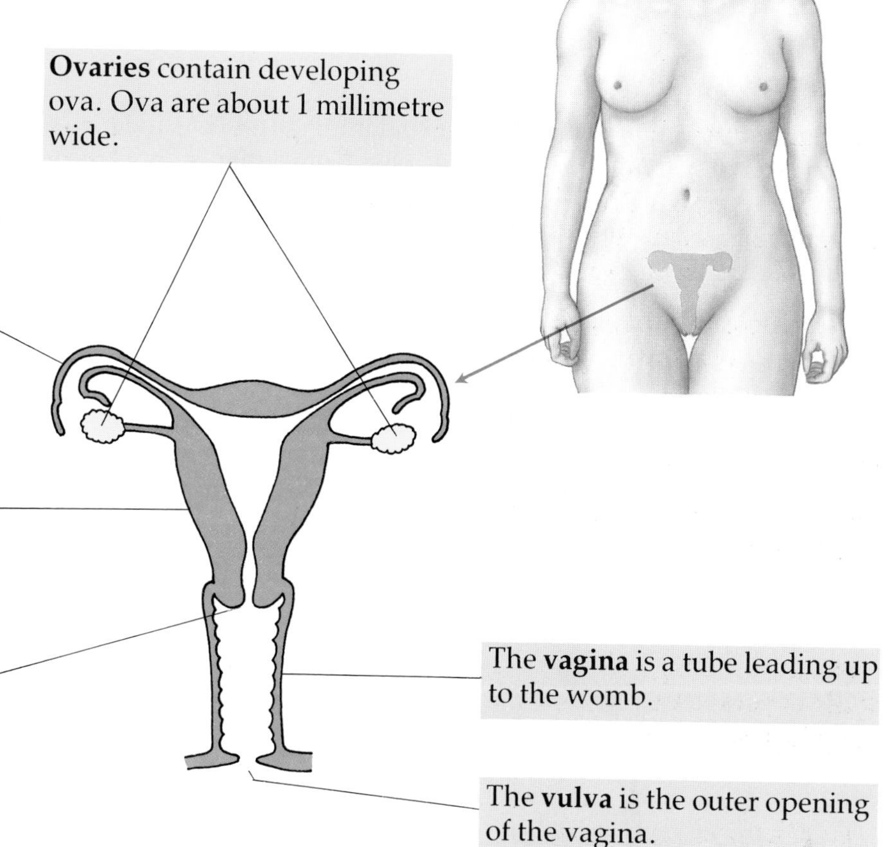

Ovaries contain developing ova. Ova are about 1 millimetre wide.

Fallopian tubes have funnel-shaped openings which catch ova as they come out of the ovaries. Ova move down these tubes to the womb.

The **womb** or **uterus** is a bag in which a fertilized ovum develops into a baby.

The **cervix** is a ring of muscle which closes the lower end of the womb.

The **vagina** is a tube leading up to the womb.

The **vulva** is the outer opening of the vagina.

Sexual intercourse

During sexual intercourse the man's penis becomes stiff and erect. He moves it backwards and forwards inside the woman's vagina. This causes semen to pump from his testes, through his penis, into the woman's body.

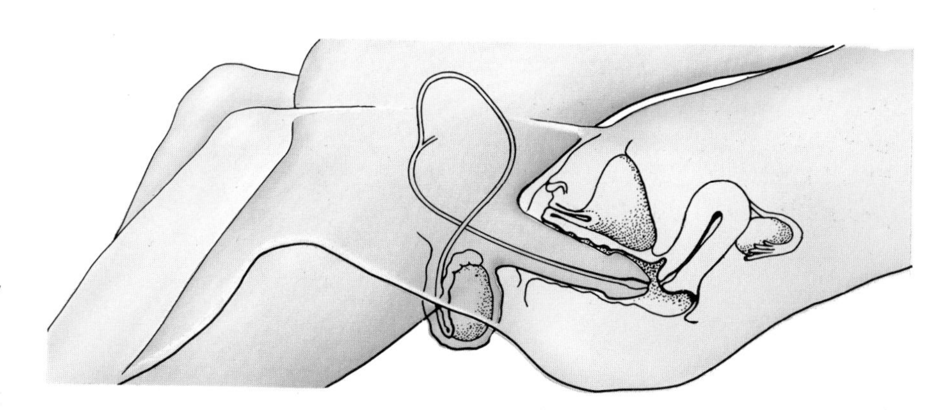

Questions

1 Which part of the body:
   a) produces sperms?
   b) produces ova?
   c) passes sperms from a man to a woman?

2 Name the liquid that contains sperms.

3 How does oestrogen affect a girl's body?

4 How does testosterone affect a boy's body?

5 What are female sex cells called?

6 How do sperms get into a woman's body to fertilize an ovum?

7 Where does a fertilized ovum develop into a baby?

8·3 Human reproduction II

The start of a new life

Human reproduction begins with sexual intercourse. During sexual intercourse millions of sperms pass from a man's penis into a woman's vagina. One of them may fertilize an ovum. This diagram shows how it all happens.

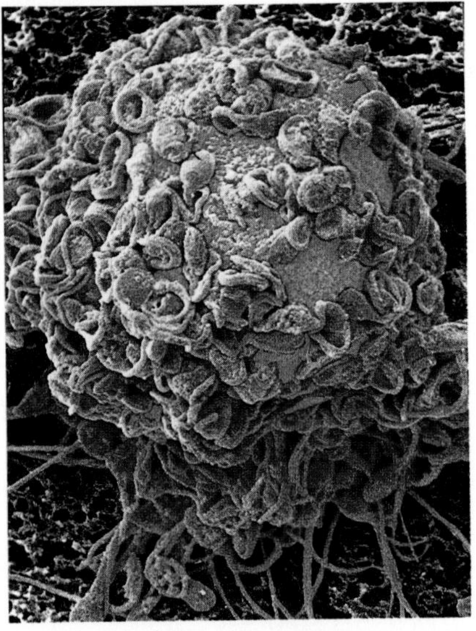

Sperms clustering round an ovum. Only one sperm will succeed in getting through the wall of the egg, to fertilize it.

1 **Ova** develop inside bubbles in an ovary. A bubble grows until it bulges from the ovary.

2 The bubble bursts squirting its ovum out of the ovary. This is called **ovulation**. The ovum is sucked into a Fallopian tube.

3 Sperms in the Fallopian tube are attracted to the ovum. One sperm burrows into the ovum. The sperm nucleus and the ovum nucleus join. This is called **fertilization**. A skin forms round the ovum to keep out the other sperms.

4 The fertilized ovum divides to make a ball of about a hundred cells. This is called an **embryo**. The embryo moves down the Fallopian tube to the womb.

5 The embryo sinks into the thick lining of the womb. This is called **implantation**. The embryo gets food and oxygen from the blood vessels in the lining of the womb. This allows it to grow into a baby.

Ovary

Lining of the womb

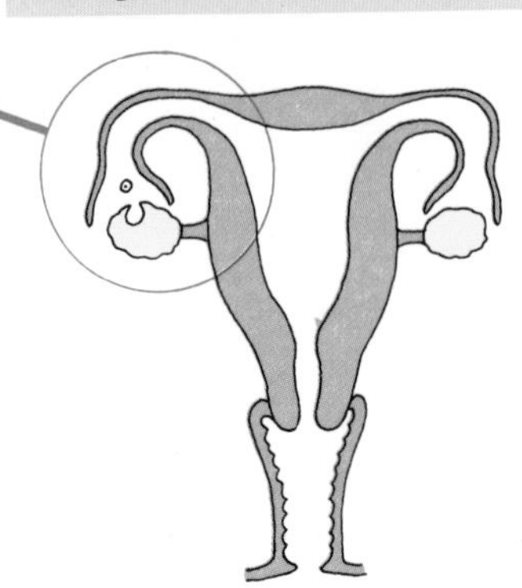

If an ovum is *not* fertilized after ovulation, it dies. But another ovum is released about a month later, as part of the **menstrual cycle**.

The menstrual cycle

Every month changes take place in a woman's body. These changes are called the **menstrual cycle**, or **monthly period**. They usually begin when a girl is between 8 and 15 years old.

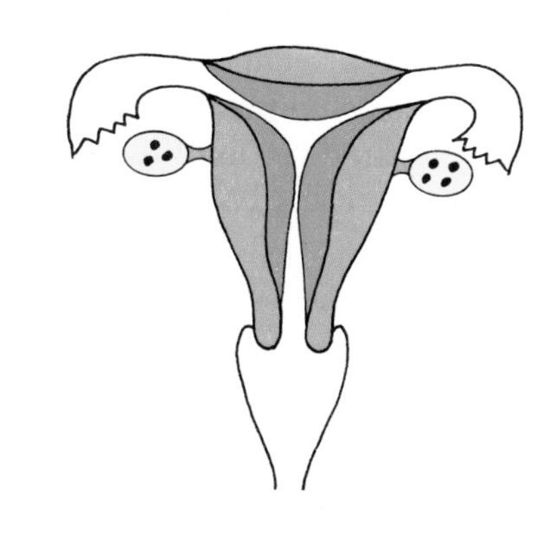

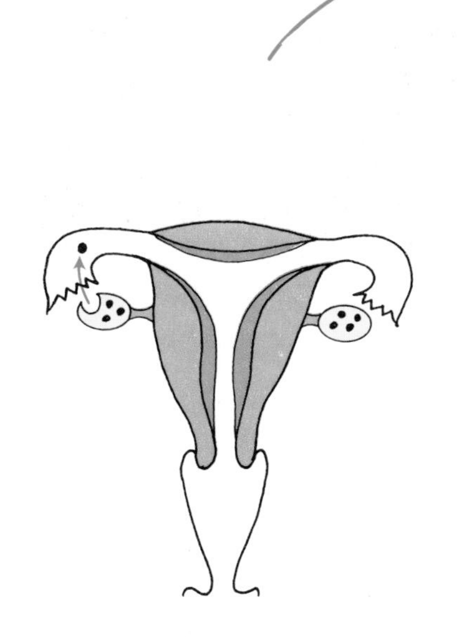

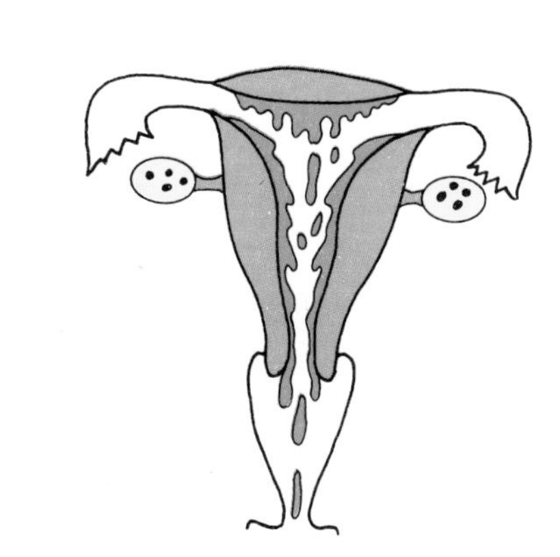

2 During the week after ovulation the womb grows a thick lining of glands and blood vessels. The womb is now ready to protect and feed a fertilized ovum.

1 Ovulation produces a new ovum about every 28 days. The ovum is sucked into a Fallopian tube. If the woman has sexual intercourse the ovum may be fertilized. The woman is then **pregnant**. But if the ovum is not fertilized it dies within a few days.

3 About 14 days after ovulation, if the woman is not pregnant, the thick lining of the womb breaks down. Blood and dead cells flow out of her vagina. This is called **menstruation**, or a **period**. Then the cycle starts again.

When a woman is between 45 and 55 years old, her periods stop. This change is called the **menopause**, or **change of life**.

Questions

1 What is ovulation? How often does it happen?

2 Where in the body does an ovum get fertilized?

3 How do sperms reach the ovum?

4 How does a fertilized ovum become an embryo?

5 Explain what implantation means.

6 During the menstrual cycle:

a) why does the womb grow a thick lining?

b) what happens when this lining breaks down?

8·4 Life before birth

A baby feeds, breathes, and gets rid of waste inside a mother's body. The mother's blood brings it food and oxygen and carries the wastes away. But the mother's blood does not mix with the baby's blood. Food, oxygen, and wastes are exchanged in the **placenta**.

The placenta is shaped like a pancake, and is attached to the wall of the womb. There are blood vessels and spaces inside it. The baby's blood flows into the blood vessels. The mother's blood flows into the spaces. Substances are exchanged through the baby's blood vessel walls.

It is important for a pregnant woman to eat healthy food, to give the baby a good start in life.

1 The baby's blood carries carbon dioxide and other wastes to the placenta, along an **artery** in the **umbilical cord**.

2 The mother's blood flows into the spaces in the **placenta**, carrying food and oxygen.

3 The food and oxygen pass into the baby's blood through the thin walls of the blood vessels. At the same time the **wastes** pass out of the baby's blood.

4 The mother's blood carries the wastes away.

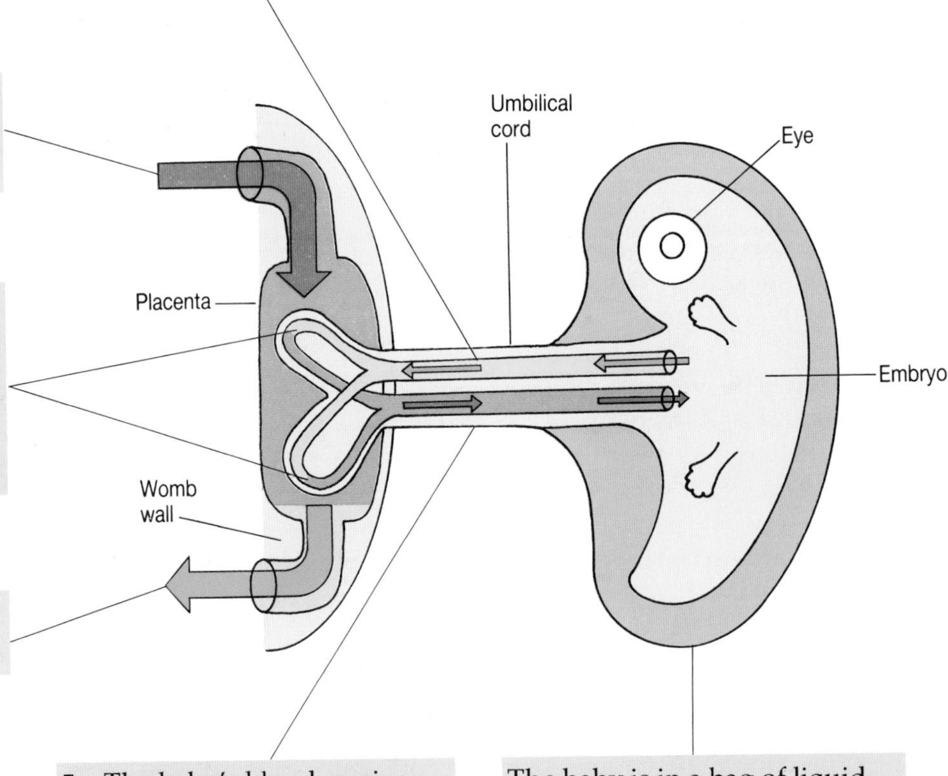

5 The baby's blood carries food and oxygen back to the baby along a **vein** in the **umbilical cord**.

The baby is in a bag of liquid called the **amnion**. This cushions it against knocks and bumps as the mother moves about.

How the baby develops

A fertilized ovum is smaller than a full-stop. In nine months, it will have grown into a baby weighing around three kilograms, and measuring about half a metre.

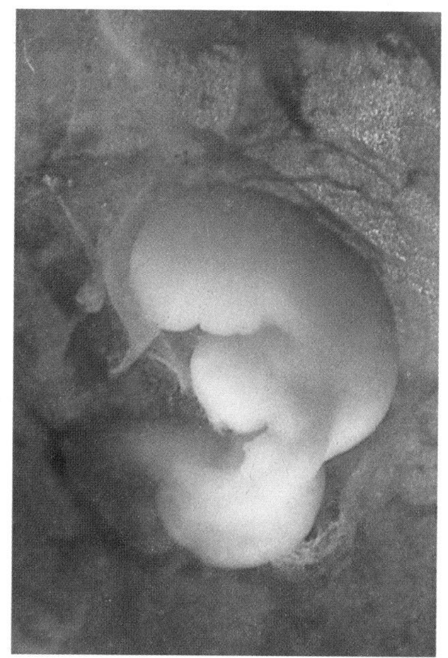

At 4–5 weeks

The embryo is about half a centimetre long. It has a bulge which will be its head, containing a developing brain, eyes, and ears. It has a tail, and lumps which will turn into arms and legs. A small heart pumps blood along an umbilical cord to the placenta.

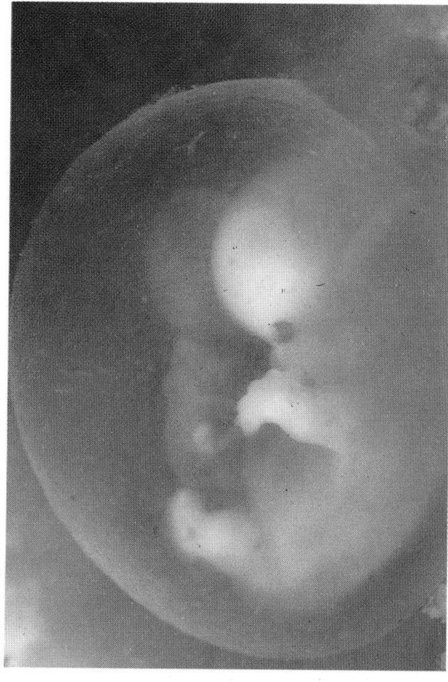

At 8–10 weeks

It is now 4 centimetres long. It has a face, and limbs with fingers and toes. Most of its organs are fully developed. From now until birth it is called a **foetus**.

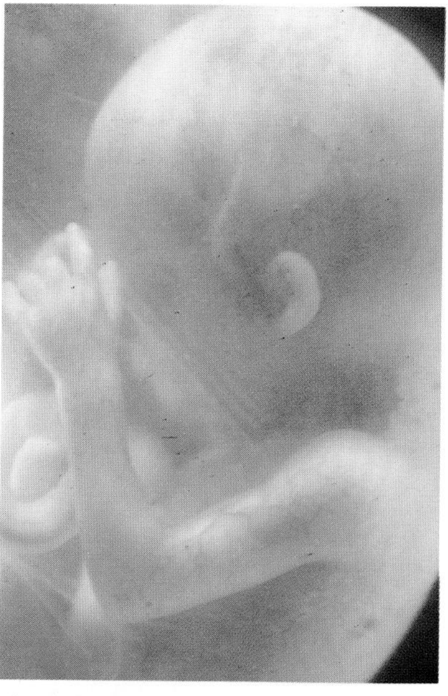

At 16 weeks

It is about 16 centimetres long. It is turning into a boy or a girl. The mother will soon feel it moving inside her.

Use the drawing on the right to name the parts of the baby in each photograph.

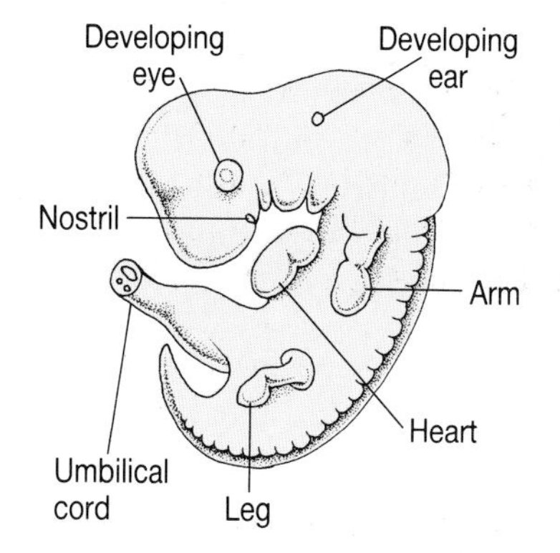

Questions

1 How does a baby get food inside the womb?

2 How does a baby get oxygen inside the womb?

3 How does a baby get rid of its wastes while inside the womb?

4 Think of a reason why the mother's blood is not allowed to mix with the baby's blood.

5 The umbilical cord has two tubes inside it. What are these tubes? What do they do?

6 What is the amnion? How does it help the baby?

7 What is a foetus?

8·5 Birth, and birth control

During pregnancy, the wall of the womb develops thick, strong muscles ready for the baby's birth.

Birth begins with **labour**. The womb muscles start to contract. After some time the contractions push the baby head-first out of the womb.

At first the womb muscles contract about every half hour. Then the contractions get faster and stronger. They push the baby's head through the cervix into the vagina.

The contractions become very strong. The baby moves to a face-down position and is pushed out of the mother's body. It breathes air for the first time.

A few minutes later more contractions push the placenta and umbilical cord out of the womb. These are called the **after-birth**.

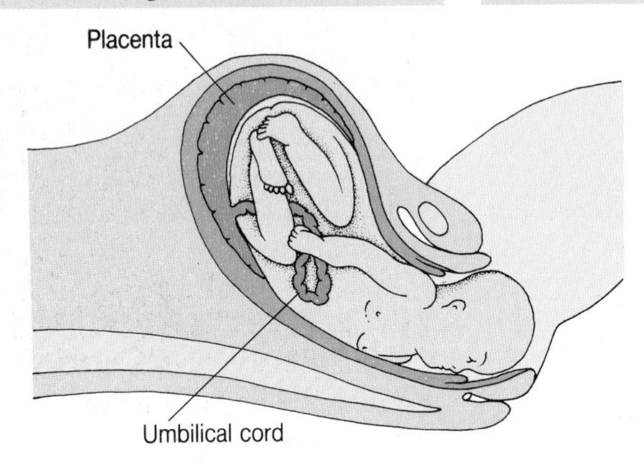

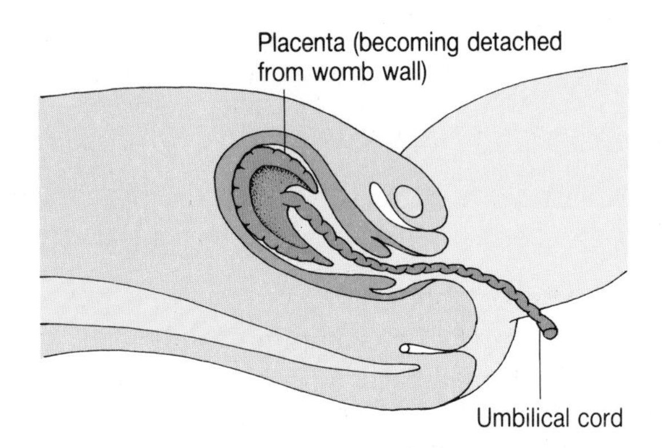

Health of the mother

When a women is pregnant, she must take care of her health.

German measles (rubella). If a mother gets rubella during the first 12 weeks of pregnancy the germs can damage her baby, causing deafness, blindness, and heart disease. Pregnant women should avoid people who have rubella. Young girls should be innoculated against the disease.

Alcohol. If a pregnant woman drinks alcohol, it can slow her baby's development, damage its brain, and cause it to be born early.

Smoking. Pregnant women who smoke have smaller babies than non-smokers. Their babies are more likely to be born dead.

Health of the baby

When the baby is born it is put on the mother's breast or stomach immediately. This helps love develop between mother and child. Mother's milk is the perfect food for babies. It is free, always ready, and pure. Mother's milk protects babies from germs. If a mother is unable to breast feed, special powdered cow's milk may be used.

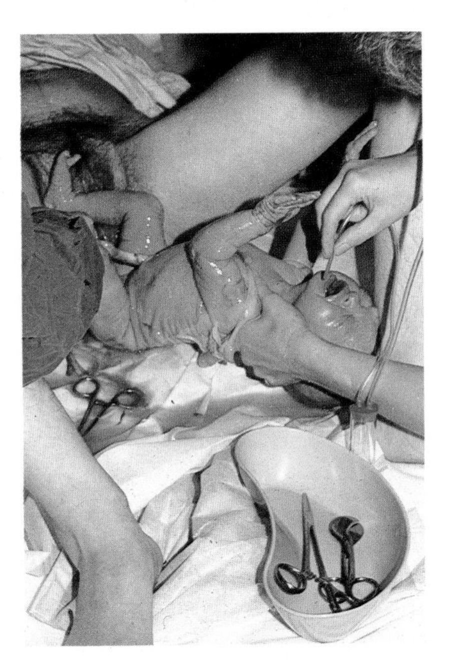

Another successful birth. The scissors is for cutting the umbilical cord.

Birth control

Birth control, or **contraception**, lets couples decide how many children they want. These are some of the methods used.

The condom. The **condom** or sheath is a thin rubber cover for the penis. It is put on before intercourse. Sperms get trapped inside it.

Condoms are quite reliable. But they are even safer if the woman uses a **spermicide**. This is a chemical which kills sperms. The woman puts a spermicide cream or jelly into her vagina before intercourse.

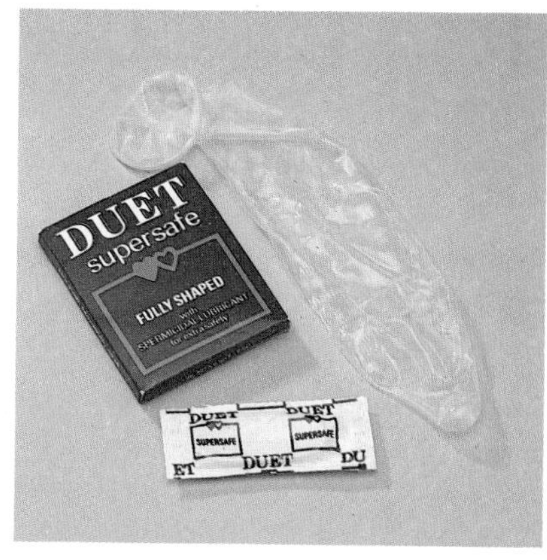

As well as being a reliable contraceptive, condoms also protect against infection.

The diaphragm. The **diaphragm** is a circle of rubber with a metal spring round it. Before intercourse, the woman smears it with spermicide. Then she puts it into her vagina, to cover the cervix. It stops sperms getting into the womb. For safety, it is left in place for at least eight hours.

Some women choose the diaphragm or cap as a contraceptive.

Contraceptive pills. Contraceptive pills contain female hormones. They do three things:

1 they prevent the ovaries from releasing ova

2 they stop an ovum moving along a Fallopian tube

3 they fill the entrance to the womb with a sticky substance so that sperms cannot swim through.

Contraceptive pills are very reliable. But they should not be used by women with liver disease or diabetes. Also some doctors think that women who smoke may develop blood clots, migraine headaches, or heart conditions, if they take the Pill. Some may also develop breast cancer.

The rhythm method. Some people feel it is wrong to use any kind of birth control except the **rhythm method**. In this method the couple avoids sexual intercourse when there is an ovum ready to be fertilized. Knowing when the ovum can be fertilized is difficult to work out. This method is not very reliable.

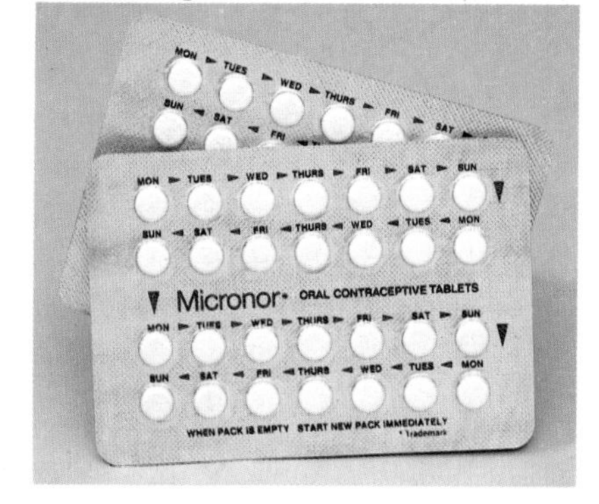

It is essential to talk to the doctor before using the Pill. It can damage the health of some women.

Questions

1 What causes a baby to be pushed out of the womb?

2 Which part of the baby normally comes out first?

3 What is the after-birth?

4 Why should young girls be innoculated against rubella?

5 Why should pregnant women avoid alcohol and smoking?

6 How does a condom prevent pregnancy?

7 What is a spermicide? Which methods of contraception are made safer by spermicides?

8 Which women might find contraceptive pills harmful?

8·6 Population

Every day, the number of people in the world increases by 200 000. So every second there are two more mouths to feed.

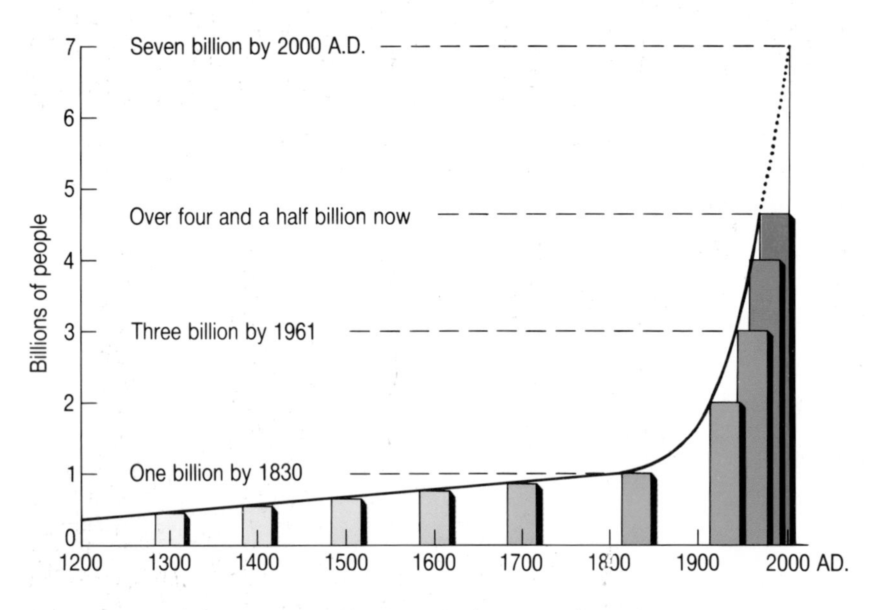

The population explosion means that the world is getting crowded. Although the Earth has enough resources for us all, still millions starve. We must take action.

Why the world's population is increasing

The world's population is increasing because:

1 better medicines and health care are helping us to fight disease

2 clean water supplies and better sewage disposal mean that diseases do not spread so easily

3 modern farming methods have given some countries more food

This means fewer babies die, and adults live longer than they used to. But in poor countries millions of people starve to death each year. Population growth is only one reason why.

Why do millions starve?

Right now there is enough food in the world for everyone. So why do millions of people starve?

1 Unequal sharing. People in rich countries build up huge stores of food, and eat far more than they need. The average European gets through enough food and other materials for 40 poor Africans.

This drawing shows the amount of food eaten in rich and poor countries each year.

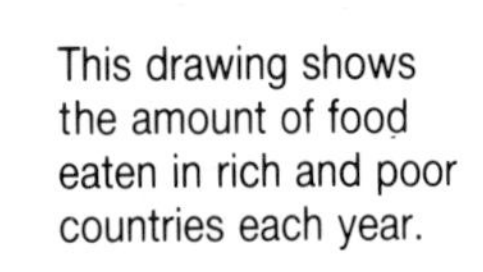

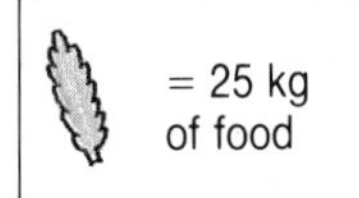

2 Growing the wrong crops. Poor countries often use their best land for tobacco, tea, coffee, and other crops they can sell to rich countries, instead of growing food for their own people.

3 Creating deserts. Tropical forests are cut down to make room for more crops. But fewer trees mean less rain. The soil is baked hard by the sun and turns into desert.

4 Population growth. People in the poorer countries tend to have the largest families. So every year it becomes more difficult for poor countries to feed, clothe, and educate everyone.

A tropical forest being cut down in Brazil, to make way for crops and cattle ranches. The exposed topsoil will soon get worn away by wind and rain.

Land is used to grow crops like coffee for sale instead of for feeding the local people.

Action needed

1 The best farm land should be used to grow food to feed the local population and tropical forests should be looked after to stop the spread of deserts.

2 It is very important to stop the growth in population. This is difficult to do, because poor families feel they need many children to help farm the land, earn money, and make them happy. Better education and health is needed for all, as well as advice on birth control.

3 Rich countries must help, but not just with food and money. They can help improve education, irrigation, and farming methods in poorer countries.

Questions

1 Use the graph on the opposite page to find out:
  a) the population of the world now
  b) the population in the year 2000, if it keeps growing at its present rate.

2 World population rose rapidly after the year 1800. Suggest some reasons for this.

3 The population is growing faster in poor countries than in rich ones. Why is this?

4 Why do some countries use their best land for crops like tobacco and coffee, rather than food?

5 Clearing tropical forests to grow crops can make things worse in the long run. Explain why.

Questions

1 Copy the following sentences and fill in the gaps.

a) In. . . . .reproduction there is only one parent.
b) *Amoeba* reproduces asexually by. . . . . .
c) Male sex cells of animals are called. . . . . .
d) They are produced by organs called. . . . . .
e) Female sex cells of animals are called. . . . . .
f) They are produced by organs called. . . . . .
g) . . . . .is the joining together of a male sex cell and a female sex cell.
h) A fertilized cell is called a. . . . . It develops into a baby.
j) A partly-developed baby is called an. . . . . .

2 **a)** Name some animals that have external fertilization.
b) Name some animals that have internal fertilization.
c) Write down four advantages of having a baby develop inside its mother's body.
d) Write down four disadvantages of having a baby develop outside its mother's body.

3 **a)** How many sperms are needed to fertilize an ovum?
b) What happens to the nucleus of a sperm and ovum during fertilization?
c) Why does a skin form around an ovum immediately after fertilization?
d) How long does it take for a fertilized human ovum to develop into a baby?
e) Where does the implantation of an embryo take place?
f) What does implantation mean?

4 **a)** What is the scientific name for German measles?
b) Describe two ways in which German measles can damage a developing baby.
c) How can girls avoid catching this disease?
d) Explain why pregnant women should not smoke.
e) Give three reasons why breast-feeding is better for babies than bottle-feeding?

5 Below is a diagram of a woman's reproductive system.

a) Name the parts labelled B to G.
b) Now match each of these descriptions to one of the labels:

where fertilization occurs;
the space where a baby develops;
ovulation occurs in this organ;
the lining of this part is shed once a month;
a baby passes through these parts during delivery.

c) Write down another name for a woman's period.
d) What happens to the lining of the womb just after ovulation? Why does this change happen?

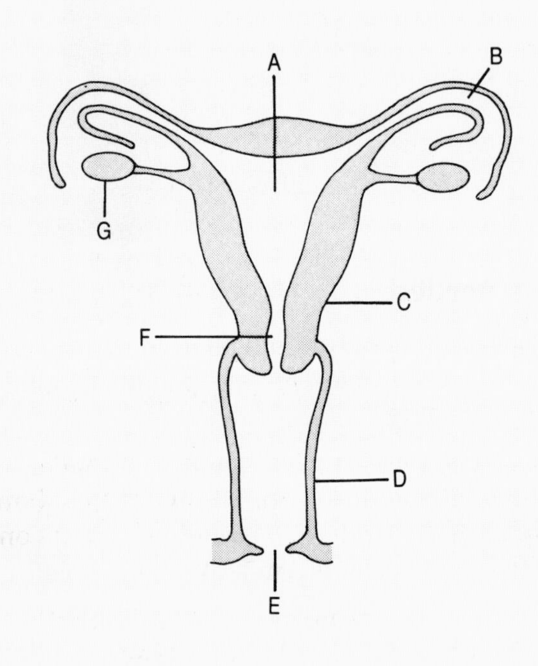

6 **a)** What is a condom?
b) What is a diaphragm?
c) Why should a woman put spermicide inside her vagina before sexual intercourse when using a diaphragm or condom?
d) What do contraceptive pills contain?
e) Describe three ways in which contraceptive pills can prevent pregnancy.

7 Look at the diagram of a man's reproductive system opposite.

a) Name the parts labelled A, B, C, and E.

b) Now match each of these descriptions to one of the labels:

passes sperms into the woman's reproductive organs during sexual intercourse;
produces the hormone testosterone;
makes the penis erect during intercourse;
carry sperms from the testes to the penis;
make the liquid part of semen.

c) List the changes which the hormone testosterone produces in boys. At about what age do these changes begin?

d) List the changes which the hormone oestrogen produces in girls. At about what age do these changes begin?

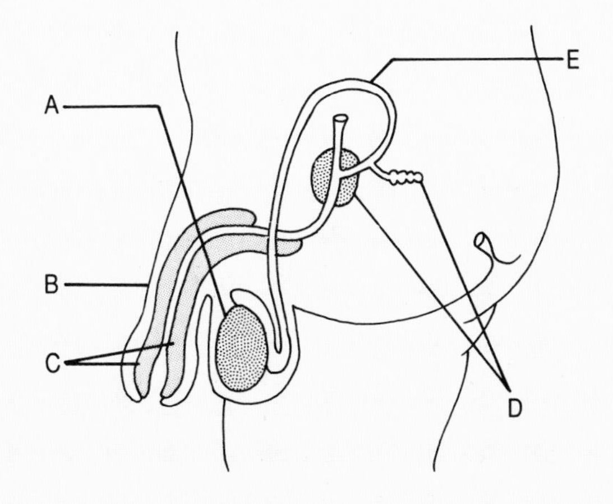

8 The drawing opposite shows a developing baby inside the womb.

a) Name the parts labelled A to D.

b) What does part C contain?

c) Name two substances which pass into the baby's blood from the mother's blood.

d) Name one substance which passes from the baby's blood into the mother's blood.

e) Where on the diagram do all these substances pass between the mother and baby?

f) What is the job of the part labelled A?

g) What does part B do during the birth of a baby?

h) A few minutes after a baby is born the after-birth is pushed out of the womb. Name the parts of the after-birth. Which letters on the diagram point to these parts.

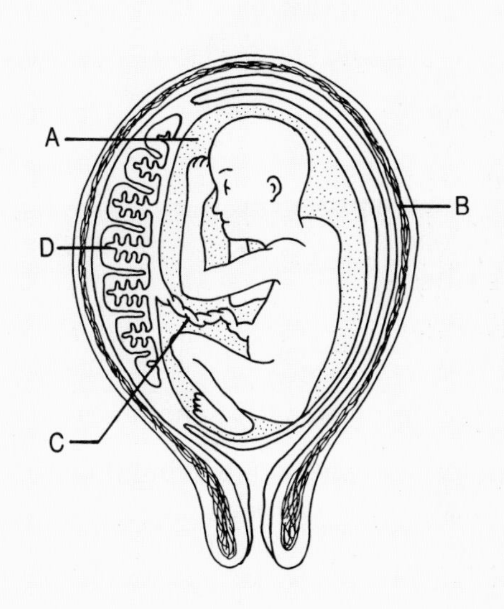

9 **a)** From the graph opposite, find out: the population of the world in 1900 and its population in 1970.

b) Describe how the population changed between those two years.

c) Write down as many reasons as you can to explain this change.

d) What problems has the growth in world population caused?

e) Give three reasons why millions of people are starving to death when there is enough food for everyone in the world?

f) Write down two actions that rich countries can take to help poor countries.

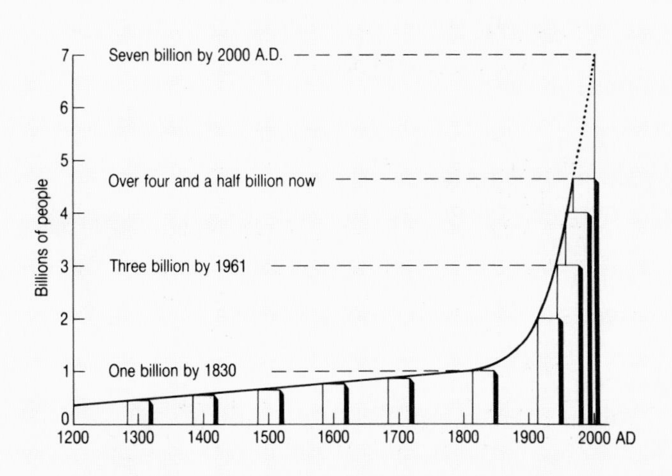

9·1 Depending on each other

Producers and consumers

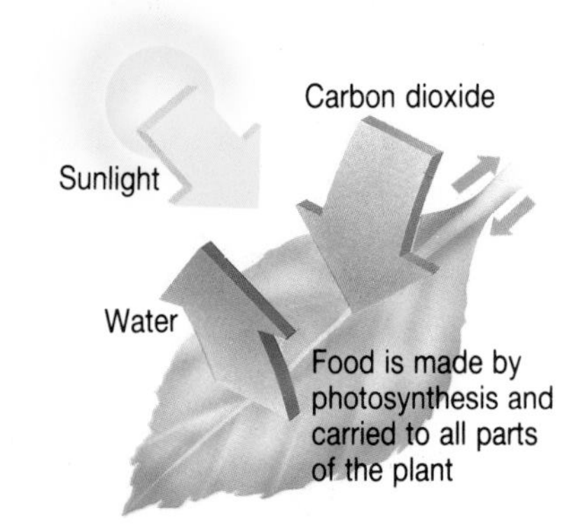

All living things need energy. They get their energy from food. Plants are called **producers** because they make their own food. They use sunlight to make food by photosynthesis.

Without plants, humans and all other animals would starve to death. This is because they *cannot* make their own food. The only way animals can obtain energy is by eating, or consuming, plants or other animals. So animals are called **consumers**.

Food chains

1 Seaweed is a **producer**. It produces food by photosynthesis.

2 Periwinkles are called **first consumers** because they eat seaweed.

3 Herring gulls are called **second consumers**, because they eat periwinkles.

Seaweed ⟶ periwinkle ⟶ herring gull.

This is an example of a food chain.

Food chains show how one living thing is the food for another. Energy passes along a food chain from producers to consumers as one member of the chain eats the next.

Food pyramids

On the right is a **food pyramid**. It shows how consumers can get larger in size, but smaller in number, as you go along the food chain.

When a moth caterpillar eats a leaf it uses up some of the energy from the leaf for itself. So there is less to pass on to a robin. That means the robin needs to eat several moth caterpillars to get enough energy. So there are fewer robins than caterpillars.

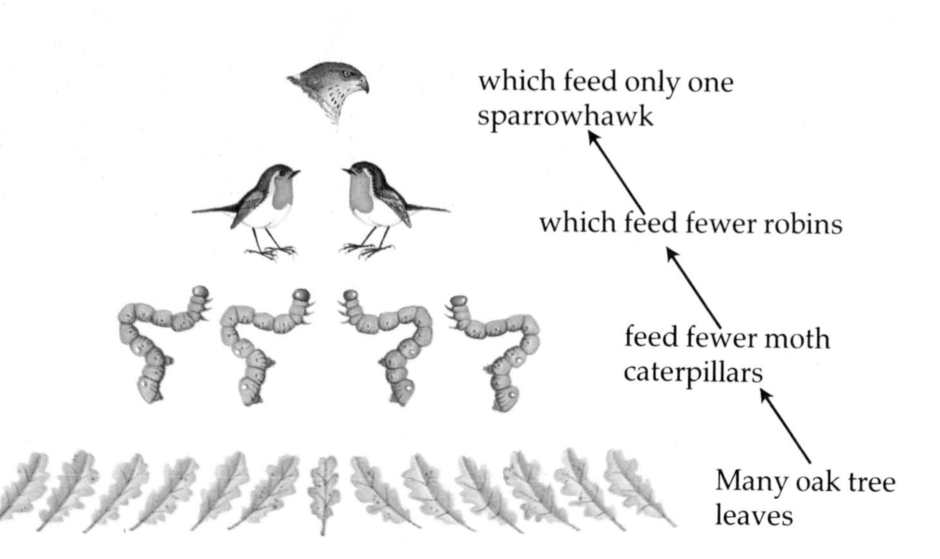

Food webs

A plant or animal usually belongs to several food chains. For example, seaweeds are eaten by limpets and winkles as well as by sea worms. In this way food chains are connected together to make **food webs**. This diagram shows a food web in the sea.

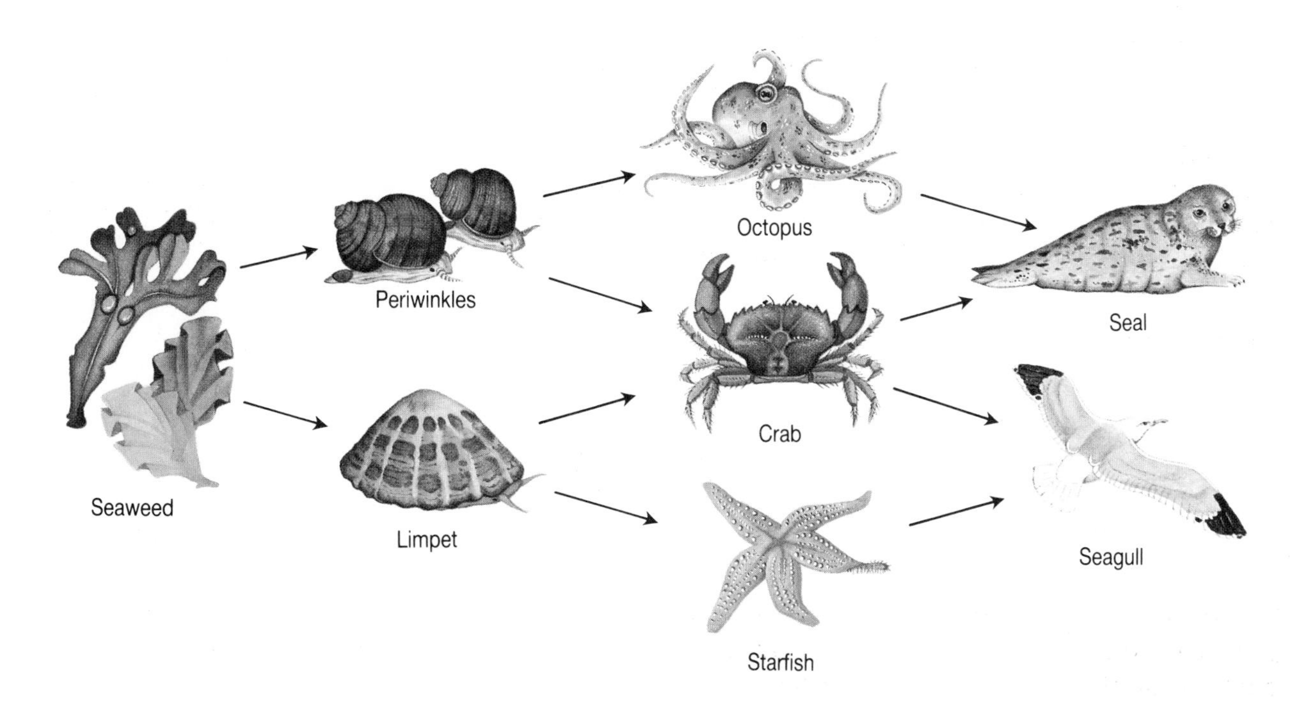

Decomposers

Many fungi and bacteria are **decomposers**. They feed on dead animals and plants. First they produce digestive juices which make the dead things rot, or **decompose** into liquid. Then they absorb the liquid.

Decomposers are important because:

1 they get rid of dead animals and plants

2 they release chemicals from dead things which soak into the soil and help to keep it fertile

Fungi feeding on a dead tree. They turn the wood into liquid.

This apple is providing a feast for millions of bacteria.

Questions

1 Why are plants called producers?

2 Put this food chain in the correct order:
slug → fox → primrose leaf → frog

3 Explain how decomposers keep soil fertile.

4 In the food web above list:
a) the producers
b) the second consumers.

5 Now write down two food chains ending in seals.

9·2 The carbon cycle

All living things need carbon. It is needed for the proteins, fats, and other substances that make up living things. The carbon comes from carbon dioxide in the air.

Plants take in carbon dioxide from the air. They use it to make food by photosynthesis. Animals then get carbon by eating plants.

The amount of carbon dioxide in the air always stays the same, because it is returned to the air as fast as plants take it in.

1 Plants and animals give out carbon dioxide when they respire.

2 Bacteria and fungi respire and give out carbon dioxide while they are decomposing the bodies of dead animals and plants.

3 Wood, coal, gas, and petrol contain carbon. When they are burned the carbon combines with oxygen to form carbon dioxide. It goes off into the air.

Wood, coal, gas and oil started off as carbon dioxide. The gas is released again when they are burned.

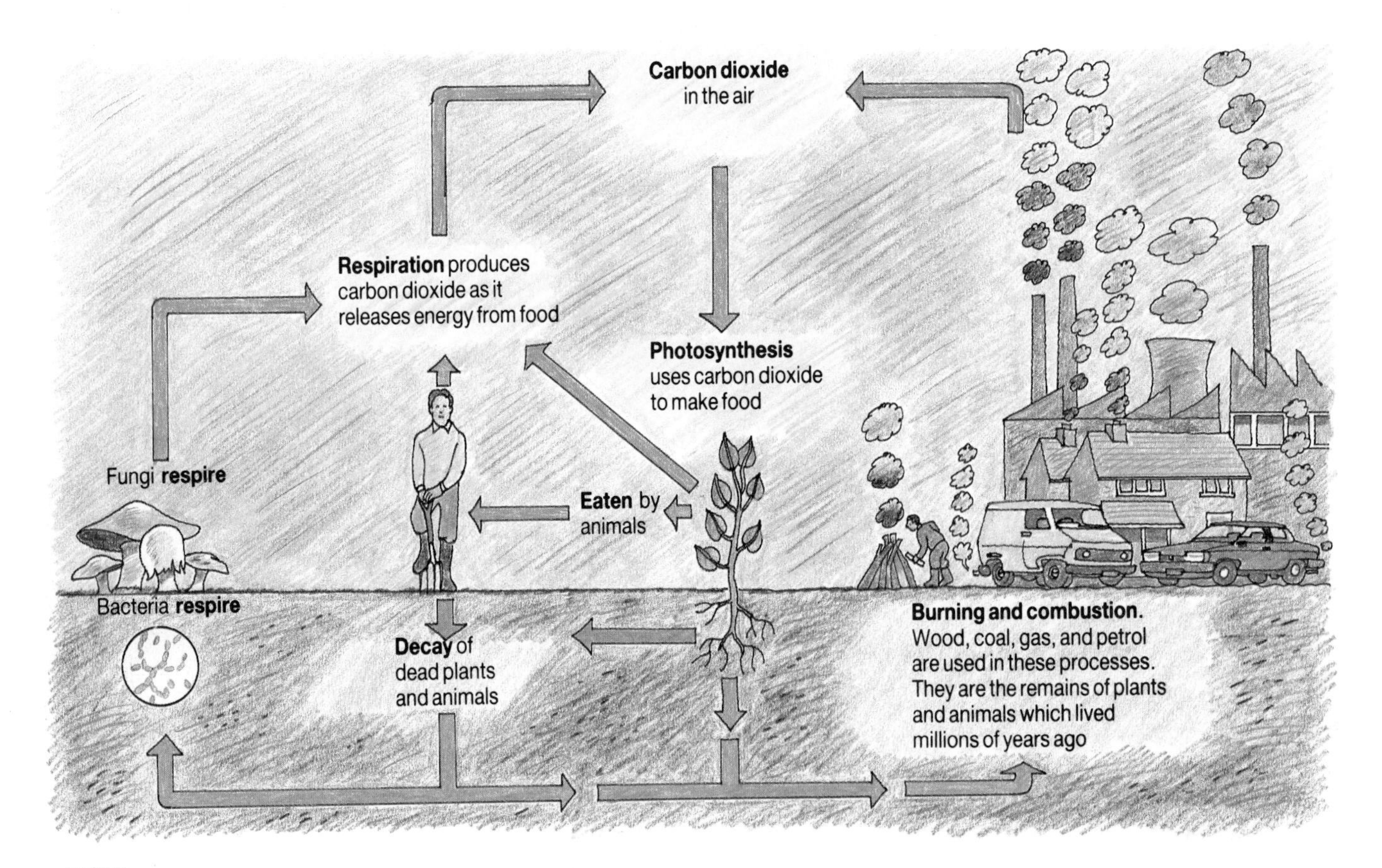

Questions

1 How does carbon get from the air into the bodies of plants and animals?

2 Explain why the amount of carbon dioxide in the air stays the same.

9·3 The nitrogen cycle

All living things need nitrogen to make proteins. Air is four-fifths nitrogen. But neither plants nor animals can take it in from the air. First it has to be changed into **nitrates**.

Then plants obtain nitrogen by taking in nitrates from soil. Animals obtain nitrogen by eating plants, or other animals.

Root nodules on a pea plant, homes for nitrogen-fixing bacteria. Because the bacteria feed them with nitrates, peas grow well even in poor soil.

How nitrogen is turned into nitrates

1 **Lightning** makes air so hot that nitrogen and oxygen combine. They make chemicals which are washed into the soil, where they form nitrates.

2 **Nitrogen-fixing bacteria** live in soil, and in lumps called **root nodules** on the roots of clover, peas, and beans. They use nitrogen from the air to make nitrates, which are released into the plants or the soil.

3 **Nitrifying bacteria** make nitrates from animal droppings, and from the bodies of dead animals and plants.

Some soil nitrates are changed back into nitrogen by **denitrifying bacteria**. These live in the waterlogged soil of ponds and marshes.

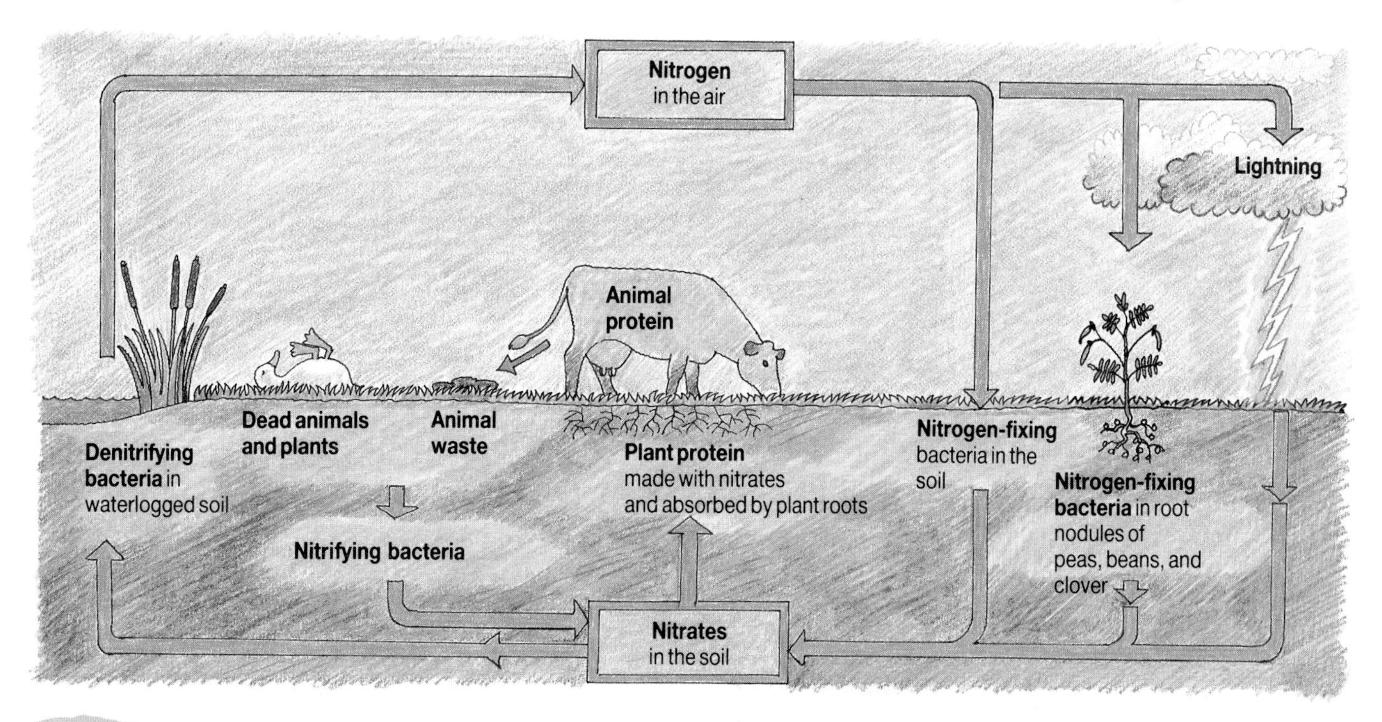

Questions

1 Why do living things need nitrogen?

2 How do plants obtain nitrogen?

3 How do animals obtain nitrogen?

4 Describe three ways in which nitrogen is changed into nitrates.

5 How are nitrates changed back into nitrogen?

9·4 Soil

All life on land depends on soil. Land plants need it for support, water, and minerals. Land animals need it because they eat plants, or other animals which eat plants.

Soil is made from small bits of **rock**, and from **humus** which is the decayed remains of dead animals and plants. The rock is broken into pieces by the weather. This is called **weathering**.

How soil is formed by weathering

1 Rock gets broken up when **rain water** soaks into it and freezes. Rain also contains weak acid which can dissolve rock and carry it away.

2 **Heat and cold** make rocks expand and contract. After a time this can make them crack.

HEAT

COLD

RAIN

First, weathering splits the rock into small pieces.

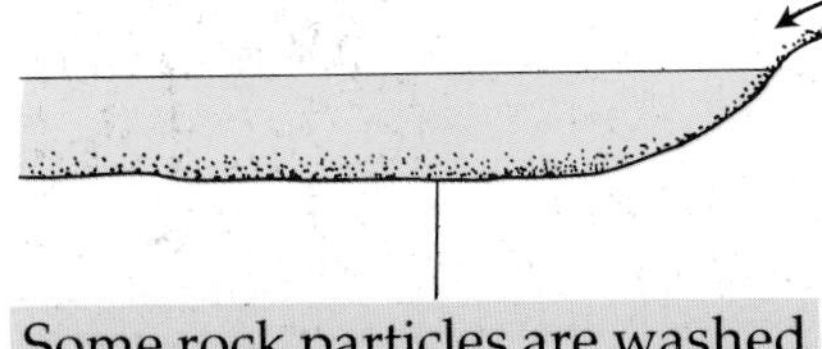

Then weathering breaks these down into smaller rock particles. Soil is formed when they mix with humus.

Some rock particles are washed into rivers and worn down into sand, clay, and silt.

What soil is made of

If you shake up soil with water it settles into different layers.

Humus floats to the top. It is important because:

1 it is rich in minerals

2 it contains nitrogen-fixing bacteria

3 it sticks rock particles together so they don't get blown or washed away

4 it keeps soil moist because it soaks up water like a sponge.

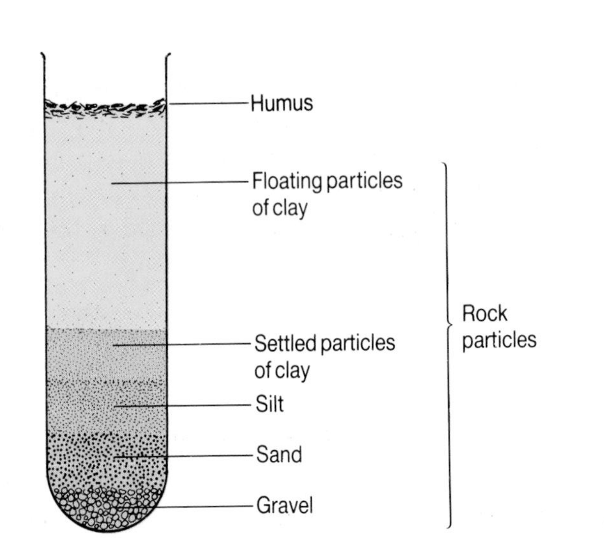

Fertile soil

The most fertile soil is called **loam**. Its rock particles are stuck together by humus into clumps called **soil crumbs**.

Soil crumbs stop minerals being washed from soil. They soak up water, and have air spaces in between them. So roots have minerals, moisture, and air. Soil crumbs also make soil easy to dig.

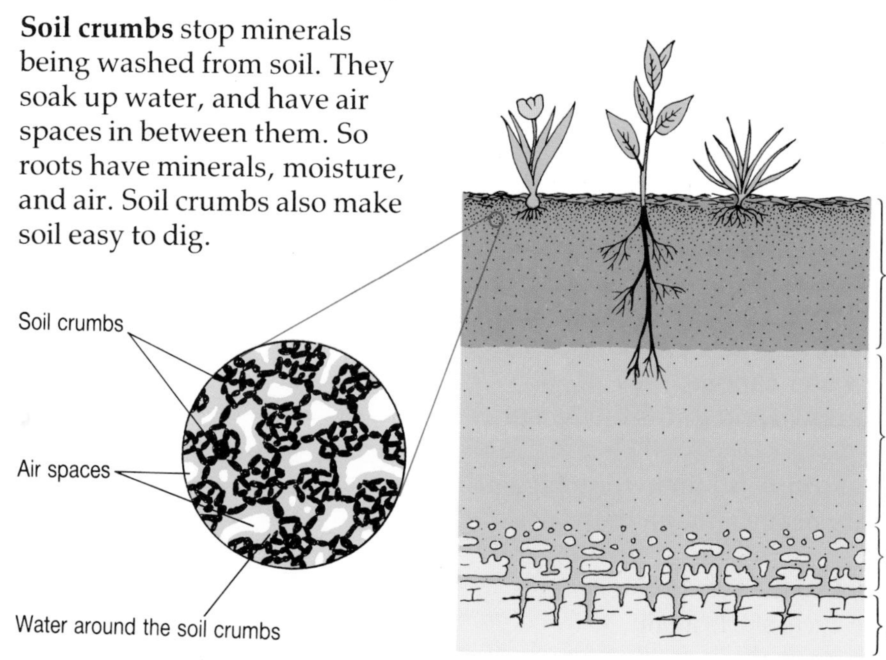

Topsoil is very fertile. Humus gives it a dark colour. It is rich in minerals, and contains many living things.

Subsoil is lighter in colour because it has no humus. It is rich in minerals but has few living things, except plant roots.

Fertilizers

Garden and farm soil must have fertilizers added, to replace the important minerals that plants use up.

Organic fertilizers include manure, well-rotted compost, dried blood, and bonemeal. These add humus as well as minerals to soil. But they work slowly because they must decompose first.

Inorganic fertilizers are chemicals made in factories. They dissolve in soil water, so plants can use them immediately. But if they are used over a long time, they harm soil by destroying soil crumbs.

Soil life

The most important living things in soil are the bacteria and earthworms. They both help to keep soil fertile. The bacteria decompose dead plants and animals. When the earthworms burrow through soil, they mix its layers together, and help air and water get into it. They also drag in leaves which rot to form humus.

An earthworm in its burrow. Earthworms eat the soil as they burrow through it. They digest some, and pass the rest out as worm casts.

Questions

1 How does weathering turn solid rock into soil?

2 What is humus?

3 Why is humus an important part of soil?

4 How can chemical fertilizers make soil less fertile?

5 How do earthworms help keep soil fertile?

9·5 The water cycle

Water is always on the move. It rises from seas, rivers, and lakes to the clouds and falls back again as rain. This is called the **water cycle.** Land plants and animals would die without the water cycle, and soil would turn to dust.

How the water cycle works

1 The sun warms seas, rivers, and lakes. The heat causes water to evaporate and turn into **water vapour**.
2 Plants and animals also give out water vapour.
3 As the water vapour rises it cools down and forms tiny water droplets. These gather as **clouds**.
4 When clouds cool down the water droplets join to form bigger and bigger drops. When the drops are heavy enough they fall as **rain**.
5 Streams and rivers take water back to the sea. Some water is taken up by plants and animals. The cycle starts again.

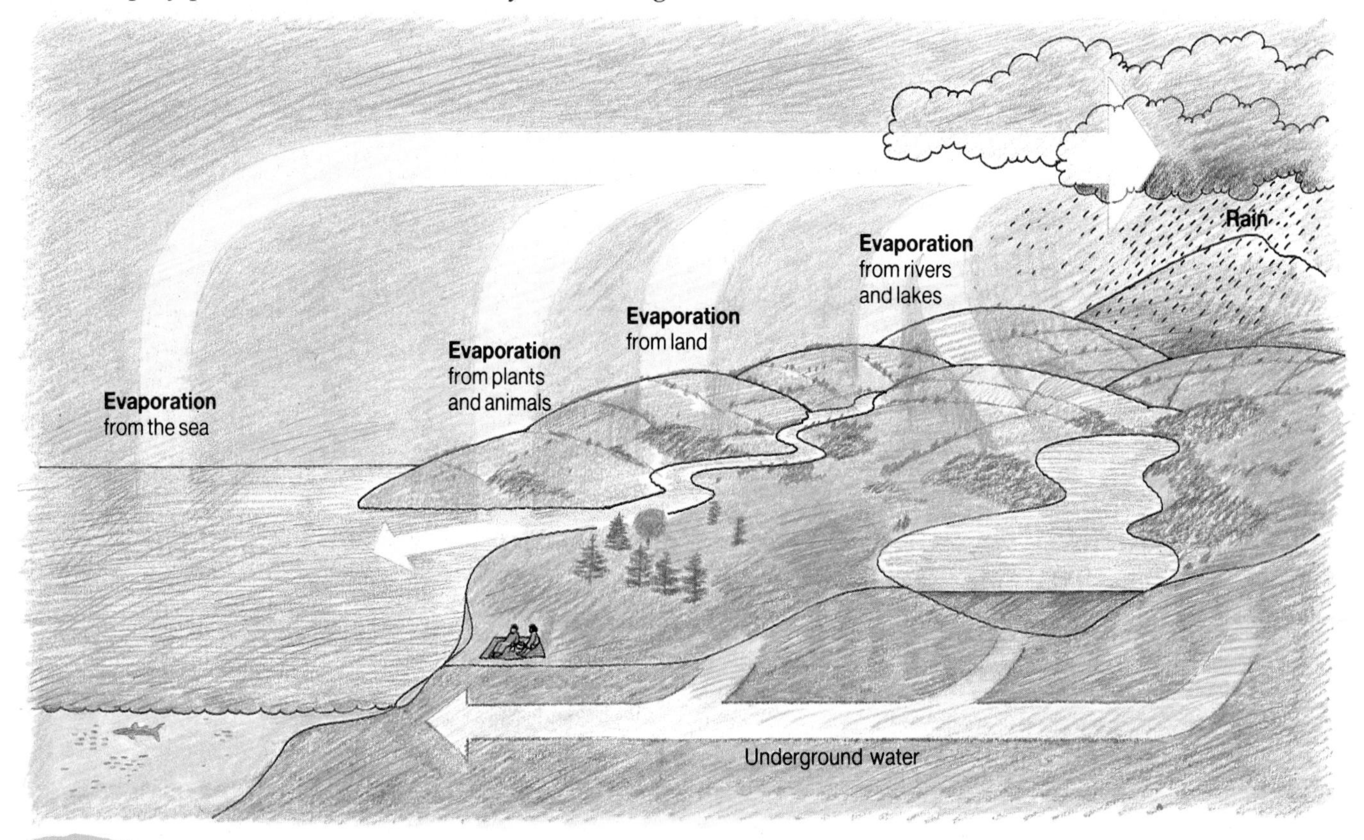

Questions

1 Why is the water cycle important?
2 What part does the sun play in the water cycle?
3 How are clouds and rain formed?
4 How does water get back to the sea?

9·6 Water pollution

The River Nishua in the USA: a dead river, polluted by waste from paper mills.

Thousands of different plants and animals live in water. Water life can be killed and our water supplies poisoned if harmful substances, called **pollutants**, get into the water.

Some water pollutants

Waste from factories may contain poisonous chemicals such as cyanide, mercury, or lead. These can pass along food chains to fish. So birds, humans, and other fish-eating animals get poisoned.

Fertilizers sprayed onto fields can soak into rivers and lakes where they cause algae to grow fast. The algae spread over the surface blocking out light from the plants below. The plants die and are decomposed by bacteria which use up all the oxygen. So fish and all other water life die too.

Pesticides are sprayed onto crops to kill insects and other pests. But they also kill wildlife in hedgerows, ponds, and streams.

Waste from farms is called **slurry**. It is a mixture of animal droppings and urine. It is sprayed onto fields as fertilizer. Because it is liquid, it can easily seep through the ground into streams and rivers, and pollute them.

Oil spilled from tankers and offshore oil rigs poisons sea birds, and clogs their feathers so they cannot fly. It also kills animals and plants that live along the sea shore, and spoils holiday beaches.

Pesticide kills harmful pests – and harmless wildlife too.

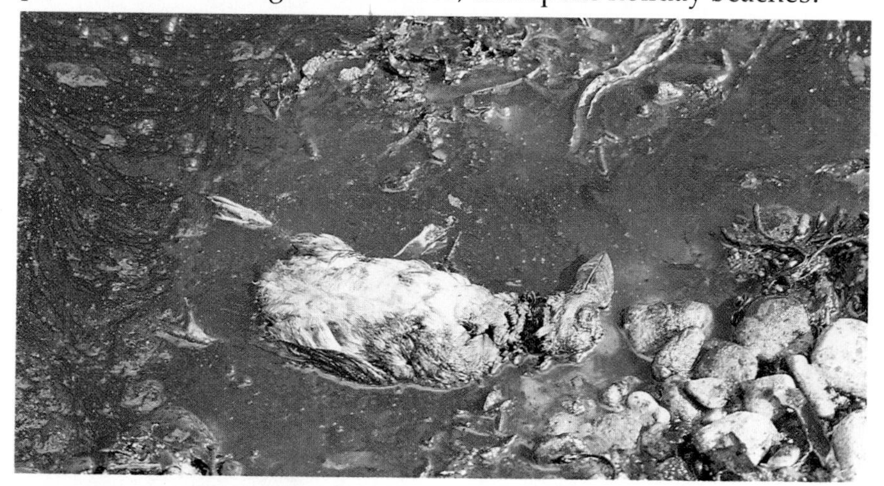

A puffin killed by oil from a stricken tanker.

The water we drink is free from pollutants. We do not poison ourselves – yet!

Questions

1 What is a pollutant? Name three pollutants.

2 How can fertilizers kill water life?

3 Describe two ways in which water pollution can harm humans.

9·7 More about pollution

Air pollution

Air is polluted by smoke, dust, and harmful gases. Most of these come from power stations, factories, cars, buses, and lorries.

Smoke contains tiny particles of carbon (soot). This blackens buildings. It covers plant leaves so that photosynthesis slows down. The smoke from car engines also contains lead, which can cause brain damage in young children.

Dust from quarries, saw mills, and asbestos factories can cause lung disease.

Harmful gases. The main ones are **sulphur dioxide** and **nitrogen oxides**. These form when coal, oil, and petrol are burned in power stations, factories, engines, and homes. They damage plant leaves, and aggravate diseases like bronchitis.

When sulphur dioxide and nitrogen oxides rise into the air they can dissolve in clouds and form acids. Then they fall back to earth as **acid rain**. This rain corrodes metal railings and bridges, and eats away stonework on buildings.

It's difficult for a leaf to photosynthesize when it's completely covered in soot!

Acid rain damage

Acid rain eats away stonework. This was once an angel.

Spruce trees in Germany, killed by acid rain.

Some pollution from cars...

...falls as acid rain.

Other kinds of pollution

Radiation. There is a risk that radiation will escape from nuclear power stations and from stored radioactive waste. Radiation can cause cancer and leukaemia.

Litter. Plastic bags and drink cans do not rot. Even paper takes a long time to rot. People who dump litter in the countryside are spoiling our beauty spots.

Noise. Noise from cars, motorcycles, aeroplanes, dogs, children, radios, and televisions can be a form of pollution. It can irritate people and cause mental depression. Prolonged loud noise from disco music and factory machinery can make people deaf.

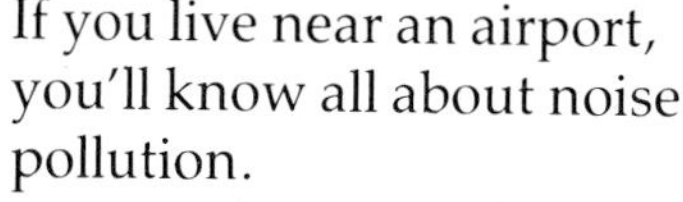

If you live near an airport, you'll know all about noise pollution.

Some ways to cut down pollution

1 By passing laws. The Clean Air Act of 1968 is an example. It turned many places into smokeless zones, where you may burn only smokeless fuel.

2 Cleaner smoke and fumes. Cleaning systems can be fitted to power station chimneys and car exhaust pipes. They stop sulphur dioxide and other pollutants from escaping into the air.

3 Cleaner fuel. Coal and oil can be treated to remove sulphur before they are burned in power stations. Petrol companies are reducing the amount of lead in petrol.

We can help cut down pollution by buying leadfree petrol...

...and joining schemes to clear-up wasteland.

Questions

1 How can soot cause damage?

2 **a)** Name two harmful gases which pollute the air.
  b) Where do these gases come from?

3 **a)** How is acid rain formed?
  b) Describe the damage which acid rain can do.

4 **a)** Why is radiation dangerous?
  b) Where could harmful radiation come from?

5 Why must you always take litter home with you after visiting the countryside?

6 When does noise become a form of pollution?

7 Describe three ways of cutting down pollution.

9·8 Conservation and ecology

We share the earth with millions of different living things. But we are destroying wildlife so fast that only the toughest species such as rats, mice, and sparrows may survive.

How wildlife is destroyed

1 Wildlife is destroyed when land is cleared for crops, farm animals, housing estates, roads, factories, mines, and quarries.
2 It is destroyed by pollution and waste tipping.
3 It is destroyed by people who hunt for fun, and to obtain things like ivory and furs which we can do without.

Grassland can have twenty or more different grasses, and scores of different flowers. Hundreds of different insects may live there. Most wild grassland has been ploughed for crops, or seeded with fast-growing grasses for cattle.

Some wildlife habitats

A **habitat** is a place where plants and animals live. Each habitat has its own collection of plants and animals.

Heathland is covered with heather, gorse, broom, billberry, lichens, mosses, and ferns. Snakes, lizards, and many birds live there. It is cleared mainly for commercial forestry, and sand and gravel quarries.

Wild woods contain oak, ash, beech, birch, shrubs, ferns, and many flowers. Most wild woods have been cleared to make way for crops or neat rows of spruce, pine, and other conifers.

Ponds and **marshes** are being drained and filled in so fast for crops and cattle that many of their plants, amphibia, birds and butterflies have become rare, and may soon disappear altogether.

How wildlife can be conserved

Over a hundred special wildlife habitats are destroyed in Britain every year. We need land to grow food, and for houses, factories, and roads. But we must not destroy all our wildlife to get it. Some places must be left untouched, for ever.

Farmers can help save wildlife by using pesticides only when it is really necessary. They can leave wild habitats untouched, and plant wild species of trees and shrubs in places where farming is difficult.

You too can help save wildlife, by following these rules:

1 Never uproot wild plants, or take birds' eggs.

2 Never leave fish hooks or lead weights where they can be swallowed by animals. The fish hooks will injure them and lead will poison them.

3 Never start fires or leave litter in the countryside.

Warning – toads crossing! Signs like these help preserve our wildlife.

To make up for the habitats we destroy, we must find ways to create new ones. An abandoned quarry makes an ideal wildlife habitat. This one has been turned into a nature park.

A wildlife garden. It contains only wild plants.

Ecology

Ecology is the study of habitats and the wildlife that lives in them. Ecologists find out which plants and animals live in each type of habitat. They study how each is adapted to live where it does. They are concerned about the environment. They advise farmers, industries and governments to consider the environment before making their development plans.

Questions

1 Describe some of the ways that habitats are being destroyed.

2 How can farmers conserve wildlife?

3 What can you do to protect wildlife?

Questions

1 a) Put each food chain in the correct order:
water snail → humans → water weed → fish
greenfly → rose leaf → ladybird
worms → dead leaves → hawk → thrush
slug → grass snake → frog → lettuce leaf
b) Name the producer in each chain.
c) Name a first consumer which is also a scavenger.
d) Most food chains have only four links. Why do you think this is?

2 Below is a diagram showing part of the carbon cycle. What do the labels A, B, and C stand for?

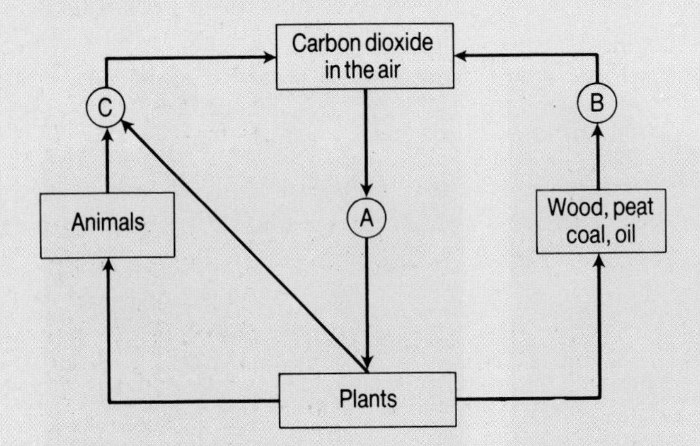

3 Below is a diagram of the nitrogen cycle.
a) How does clover help to convert nitrogen gas into soil nitrates (label A)?
b) Name the soil bacteria which convert nitrogen gas into nitrates (label B).
c) Describe one other way in which nitrogen gas is converted into nitrates (label C).
d) How are animal proteins converted into nitrates (label D)?
e) How are some soil nitrates converted back into nitrogen gas (label E)?

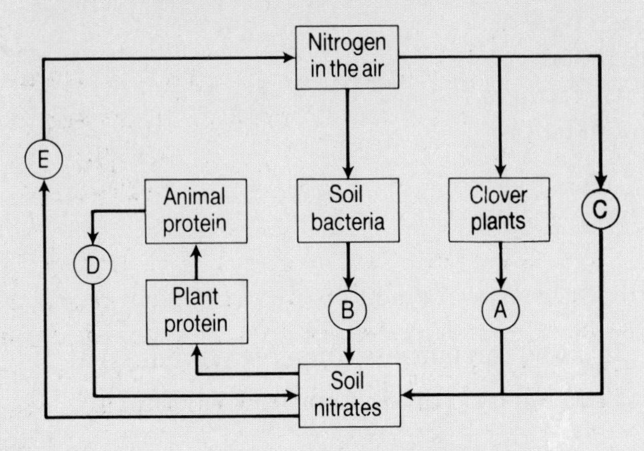

4 Below is a map of a river.
a) Fishermen catch more fish at point A than at point C?
b) Many trees in the conifer plantation are dying. Why do you think this is?
c) One day the sewage treatment works broke down, and untreated sewage entered the river. What happened to the amount of oxygen in the water at point B? Explain your answer. What effect would this have on water animals?
d) When the farmer puts fertilizer on his crops, the water weeds at point C grow bigger. Explain why.
e) Local fish-eating birds get sick when the farmer sprays his crops with pesticide. Explain why.

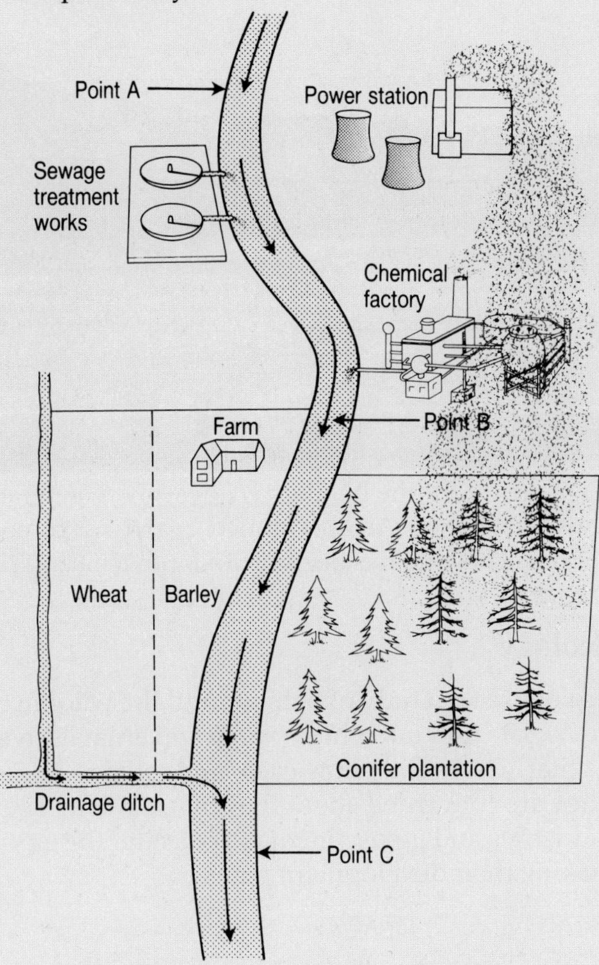

5 In the autumn, the leaves fall off trees. But chemicals from the leaves may be taken back into a tree through its roots, and used for new leaves. Explain how this happens.

1 Studying habitats

You don't have to travel long distances to find interesting habitats.

a) Grassland. A roadside verge, or even a playing field will contain many different animals and plants. Choose a suitable grassy place, and mark off a square of side 1 metre using string and pegs. Now make a list of all the plants growing within your square. Repeat for other squares. Include some squares of tall grass, if you can, and list the animals as well as the plants you find in them.

b) Hedgerows. Count the number of different *woody* plants (bushes and trees) in a 30 metre length of hedge. Multiply the result by 100. This gives you the probable age of the hedge in years.

c) Compost heaps. Take a sample of compost from the top of a compost heap. Now dig into the heap and take samples from different levels. Spread the samples on newspaper, and collect the animals from them in labelled specimen tubes. Are different animals found at different levels in a compost heap? If so, try to find out why, by measuring the temperature and water content of the heap at different levels.

d) Plants. A single plant can be a habitat for many different animals. Dig up ten dandelions (or thistles if you have thick gloves). How many different animals can you find hiding inside flowers and seed heads, on the outside of the plant, or burrowing inside the stem, roots, and leaves? Look at the animals through a hand lens. Can you spot anything that makes them suited to life in one part of a plant?

2 Making habitats

a) An aquarium. Spread some washed sand and gravel on the bottom of an aquarium. Now fill the aquarium with clean pond or river water, and plant as many different types of pond weed as you can find. Put in water insects (except the larvae of dragonflies and water beetles), and minnows. Keep the aquarium in a cool shady place. If it is in direct sunlight, it will get too hot, and turn green.

Dragonfly and beetle larvae can be kept in separate jars and fed on pieces of meat attached to a thread, which should be changed daily.

b) Log piles. Make a pile of logs somewhere they can be left undisturbed. In a few months they will be a home for centipedes, millipedes, beetles, slugs, snails, woodlice, and many other animals.

c) Insect gardens. Try growing these plants which attract butterflies and other insects: buddleia, golden rod, wallflower, Sweet William, candytuft, ageratum, geranium, thyme, phlox, verbena, and alyssum. You could also have a wildflower garden including thistles, ragwort, nettles, clover, and dandelion. Keep a record of the kinds of flowers which different insects visit.

3 Adaptation

Catch a butterfly, a housefly, a beetle, a bee, and a mosquito or gnat. You can slow them down by putting them in a specimen tube in a refrigerator for five minutes. Look at the mouth parts of each insect through a hand lens. How are these parts suited to the way the insect feeds?

4 Air pollution

Here is a simple way to observe air pollution. First, get several blocks of wood, all the same size. Paint each block white. Now put the blocks outside, in various places in your neighbourhood. For example you could put one near a busy street or a factory, and one in a park or country area. After a few months collect the blocks, and compare the amounts of dirt on them. Which place has the dirtiest air in your neighbourhood? Try to find out where the dirt is coming from.

5 Water pollution

Look for signs of pollution in local ponds, canals, rivers, streams, and beaches. If you find dead fish and birds, or oil or chemicals coming from a factory, report it to the police, or your Regional Water Authority.

If a stream contains mayfly nymphs and stonefly nymphs it is clean. If these nymphs are absent, but you can find the water louse and caddis fly larvae, it could be slightly polluted. If you find blood-red worms, or nothing at all, it is badly polluted.

10·1 Germs, disease and infection

Germs are tiny living things that cause disease. All viruses are germs. Some bacteria and fungi are also germs.

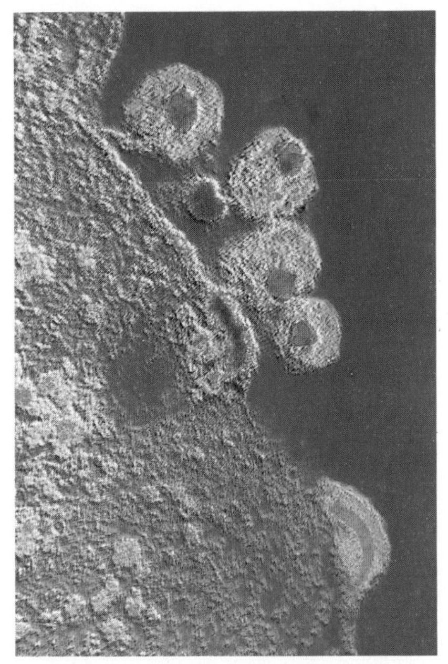

A photo of AIDS viruses, taken with false colour to show them up. The green and red blobs are the viruses. One is popping out of a white blood cell which it has just destroyed.

Viruses

Viruses are so small that you cannot see them under an ordinary microscope.

Viruses don't respire, feed, grow, or move. They just reproduce. They can only do this inside the cells of another organism. For example, if you breathe in the viruses that cause the common cold they enter cells in your nose and throat. They turn these cells into virus factories which soon infect other cells.

Chicken pox, influenza, and measles are also caused by viruses.

Bacteria

Bacteria are larger than viruses, but still very small. Unlike viruses they feed, move, and respire, as well as reproduce. Bacteria which live as germs harm living things in two main ways.

1 They destroy living tissue. Tuberculosis is a disease caused by bacteria which destroy lung tissue.

2 They produce poisons, called **toxins**. Food poisoning is caused by bacteria which release toxins into the digestive system. Boils, whooping cough, and venereal disease are caused by bacteria.

Fungi

Some fungi cause diseases in humans. For example, **ringworm fungi** cause a disease called **ringworm**. This shows up as red rings on the scalp and on the groin. A similar fungus attacks the soft skin between the toes, causing **athlete's foot**.

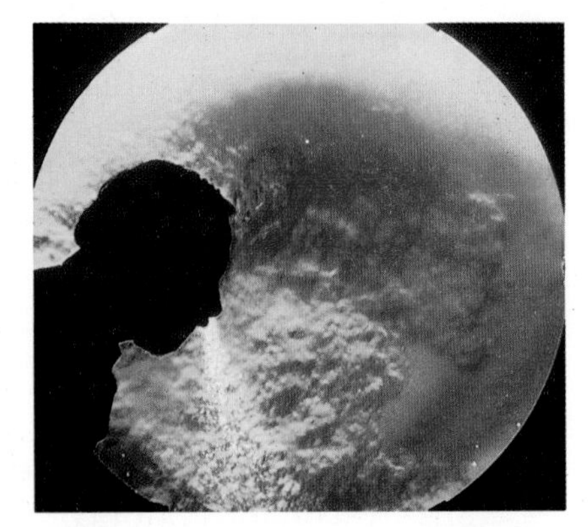

A special camera was used for this photo of a man sneezing. Note the air and droplets of mucus exploding from his nose and mouth. Can you see how coughs and sneezes spread germs?

The spread of infection

These are some of the ways you can be infected with germs:

1 By touching infected people, or things they have used such as towels, combs, and cups. Athlete's foot can be caught by walking on wet floors or mats used by infected people. Chicken pox and measles are spread by touching infected people.

2 By breathing in germs from infected people, especially when they cough or sneeze near you. So you should always cough and sneeze into a handkerchief. Colds, influenza, pneumonia, and whooping cough are spread by coughs and sneezes.

3 From infected food and drink. Food and drink can be infected with germs by coughs and sneezes, dirty hands, flies, mice, and pet animals. Infected food and drink cause food poisoning and dysentery.

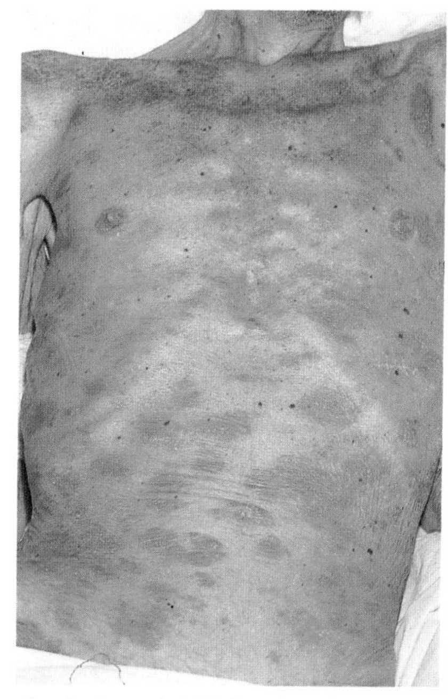

A victim of AIDS. The disease cannot be cured and so must be prevented from spreading. Wearing condoms during sex stops infection. People who use hypodermic needles should use sterilized ones.

Preventing infection

These are some ways you can avoid being infected with germs.

1 Keep yourself clean.
  a) Wash your hair regularly. If you have dandruff or lice, use a special shampoo until it clears up.
  b) Wash your face regularly.
  c) Bathe or shower regularly, especially in hot weather.
  d) Clean your teeth at least once a day.

2 Change underwear and socks daily.

3 Boil water you think may be infected with germs.

4 Cook food thoroughly and eat it straight away. If you want to store it, keep it in the fridge.

Sexually Transmitted Diseases (STD)

Gonorrhoea, syphilis and herpes are all infections that are spread by sexual intercourse. The AIDS virus can also be spread by sexual intercourse, or by an infected person's blood getting into someone else's blood.

The risk of catching sexually transmitted diseases is greater the more sexual partners you have. Using a condom during intercourse can give some protection against most of them. The best protection is not to have intercourse with an infected person.

Questions

1 List all the ways germs are being spread in the drawing above.

2 Why should you never use a stranger's towel?

3 Why should you sneeze into a handkerchief?

4 Why should you regularly wash between your toes?

5 Why must you cook food thoroughly?

10·2 Smoking and ill-health

Smoking kills. In Britain nearly 100 000 people die each year from diseases caused by smoking. This is twelve times the number killed in road accidents.

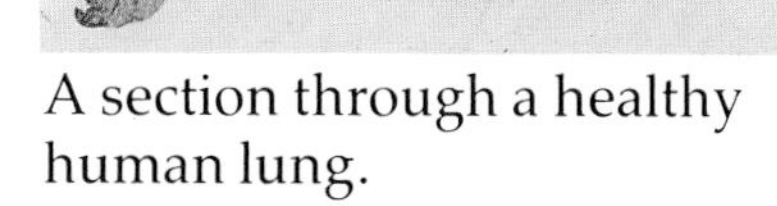

A section through a healthy human lung.

Tobacco smoke is poison

Tobacco smoke contains about 1000 chemicals. Many of them are harmful. Here are some of the poisonous ones.

Nicotine is a poisonous drug. It damages the heart, blood vessels, and nerves. Smokers get addicted to it, which is why they find it hard to give up smoking.

Tar forms in the lungs when tobacco smoke cools. The tar contains 17 chemicals that are known to cause cancer in animals.

Carbon monoxide is a poisonous gas which stops blood carrying oxygen round the body.

Other poisonous gases in tobacco smoke are **hydrogen cyanide**, **ammonia**, and **butane**. These irritate the lungs and air passages, making smokers cough.

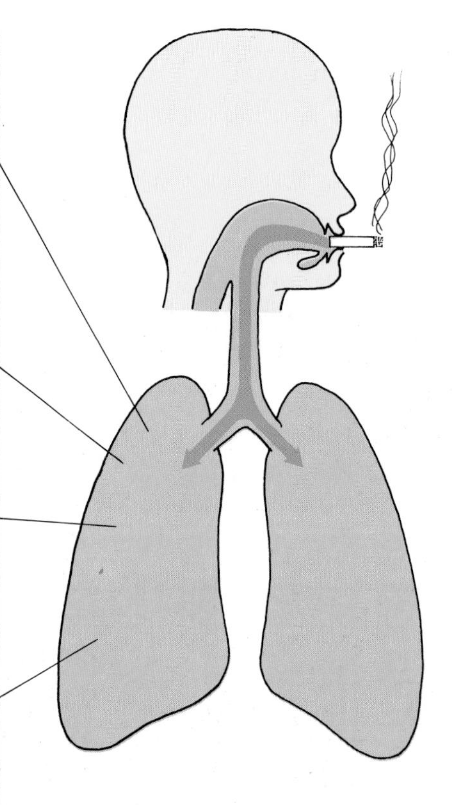

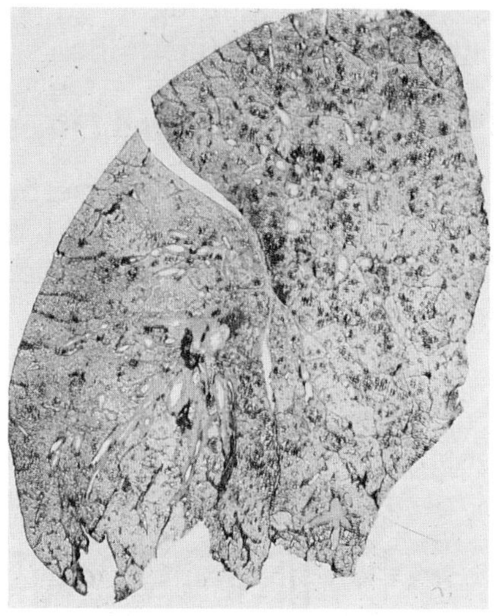

A section through the lung of a smoker. The spots are deposits of tar. They prevent the lungs from doing their job properly.

Diseases caused by smoking

Heart disease is three times more common among smokers than among non-smokers.

Emphysema is a disease in which lung tissue is destroyed by the chemicals in tobacco smoke. The lungs develop large holes which blow up like balloons. Breathing becomes very difficult.

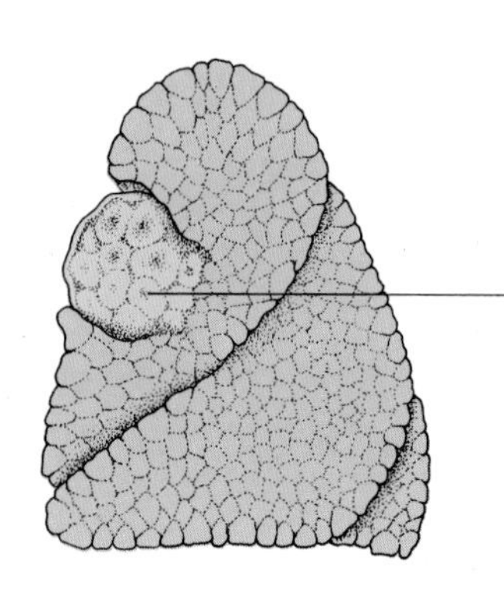

The swelling on this lung is due to emphysema, caused by heavy smoking.

Cancers of the lungs, mouth, gullet, and bladder are caused by chemicals called **carcinogens** in tobacco smoke.

Nine out of ten people who die of lung cancer are smokers.

Bronchitis is mostly a smoker's disease. Smoke irritates the air passages to the lungs, making them swollen and sore. Tiny hairs called cilia usually keep these passages clear. But the cilia stop working, so the lungs fill with mucus.

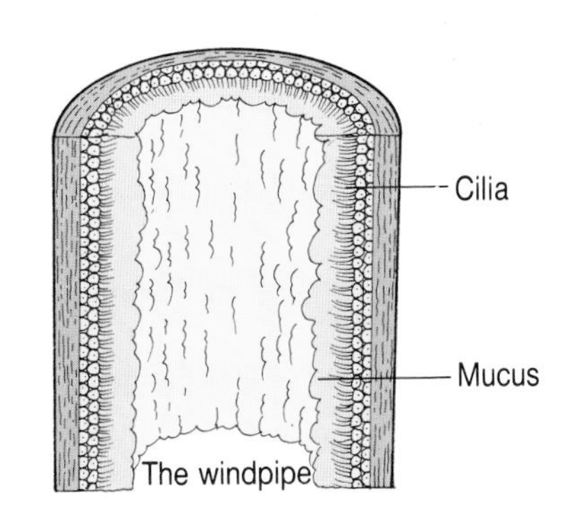

The cilia which keep your lungs and air passages clean stop working when you smoke.

Smokers get up other people's noses

Many non-smokers find tobacco smoke very unpleasant. It makes eyes sting, and can cause sore throats and headaches. It can irritate babies and people with hayfever and asthma.

Non-smokers who work or live with smokers can suffer lung damage and other smokers' diseases.

Pregnant mothers who smoke have smaller babies than non-smoking mothers. They also run the risk of a more difficult birth.

Smokers smell. Their breath, hair, clothes, and homes have a horrible smell caused by the ammonia, tar, and other chemicals in smoke.

Questions

1. Work out how much it costs to buy 20 cigarettes a day for a year. List the things you could buy if you saved this amount by giving up smoking.
2. Which chemical in smoke is addictive?
3. What harm does carbon monoxide do?
4. What is emphysema?
5. List three other diseases linked with smoking.
6. Why is it bad for pregnant mothers to smoke?

10·3 Health warning

Most people don't drink too much, or sniff glue, or take drugs like heroin and cocaine. But everyone should know the harm these things do.

Drugs and their effects

Heroin is made from opium poppies. It gives a feeling of power and contentment. But when this wears off, users become anxious and depressed. So they take more heroin to feel better. Soon they feel they cannot live without it. They are now heroin addicts.

Cocaine is made from the leaves of the coca bush. At first it gives a feeling of energy and strength. But after about 20 minutes users feel confused and anxious. They are unable to sleep for some time.

LSD is made from a fungus. Users may see things around them changing in colour, shape, and size. Or they may see things which are not there (hallucinations). Sometimes the hallucinations are very unpleasant, and are called **bad trips**. They can cause fear, depression, and mental illness.

Cannabis is made from the leaves and resin of the Indian hemp plant. Users seem to see, hear, feel and think more clearly. But they can become confused and do reckless and dangerous things.

Solvent and glue sniffers breathe in fumes from things like glue, paint, and cleaning fluid. They get a floating feeling and hallucinations. But the fumes get into the blood and damage the heart. They also cause sickness, depression, and bad facial acne.

Infections. Some drugs are taken by injection. Dirty needles can cause sores, abscesses, blood poisoning, and jaundice. AIDS can be passed from one person to another if needles are shared.

Addiction. Drugs can cause addiction. People feel they cannot live without the drug. They may turn to crime to get money for it. Once you are addicted to a drug, giving up becomes a difficult and painful experience. People begin taking drugs to feel good, they end up having to take drugs to stop feeling very bad. Drug addicts cause themselves and their families great unhappiness and stress.

Addicts can get help from Family Doctors who sometimes recommend hospital treatment. Support groups exist to help addicts and their families cope, and they may suggest ways of curing addiction.

The risks of taking drugs

Overdoses. A drug overdose can cause unconsciousness, damage to the heart and other organs, and even death. It is easy to take an overdose by accident, because it is hard to tell how strong a drug is, or how much to take. Drugs are far more dangerous when they are mixed. Their effects are multiplied, so a much lower dose can kill.

Mental harm. A drug user may suffer strong feelings of anger and fear, or terrifying hallucinations. This can cause mental damage.

Changes in behaviour. Drug users can become irritable, and lose interest in their friends, hobbies, and work. They may stop looking after themselves.

Accidents. People taking drugs run the risk of accidents, because they get confused. They can fall into water and drown, or walk in front of traffic, or fall out of windows. Glue sniffers can suffocate while their mouths are covered by plastic bags, or choke to death on their own vomit.

Alcohol

All alcoholic drinks contain a chemical called ethanol. This is what makes people drunk.

1 Ethanol is carried to the brain by the blood. From there it affects all the nervous system. A little ethanol makes people feel pleasant and relaxed. But if they keep drinking they feel dizzy and cannot walk straight. Speech becomes slurred and they begin to see double. They may become quarrelsome. Eventually they cannot stand up. If they drink more they become unconscious, and may even die.

2 Even a little alcohol causes people to make mistakes in things like typing and driving. It is very dangerous to drink and drive.

3 The effects of ethanol can last for hours. It takes about an hour for the body to get rid of the ethanol in just half a pint of beer.

4 Heavy drinking over a long period causes the brain to shrink. It can also cause stomach ulcers, cancers of the digestive system, and liver and heart disease.

5 Some people become addicted to alcohol. They turn into alcoholics.

6 If a woman is pregnant, ethanol in her blood is carried to the baby. It can damage the baby's brain and heart, and slow its growth.

Questions

1 Some drugs, and alcohol, are addictive. What does this mean?

2 Write down ten things you could say to convince a friend not to take drugs.

3 List the changes which take place in the body of a heavy drinker.

4 What changes in a person's behaviour can tell you that he or she may be taking drugs?

Questions

1 This diagram shows how bacteria are affected by temperature.

a) Between which temperatures do bacteria multiply the fastest?

b) The temperature inside the main part of a refrigerator is about −4°C. Why is it unwise to keep foods, like meat, in this part of a refrigerator for more than three days?

c) The temperature in a domestic freezer is about −20°C. Why is it unwise to keep food in a freezer for more than three months?

d) Bacteria reproduce by dividing in two. If they have food and warmth they do this about every 30 minutes. If just one 'food-poison' bacterium landed on a piece of meat, and reproduced at this rate, how many would there be after 24 hours?

e) It is dangerous to cook a large piece of frozen meat before it has thawed out properly. Explain why.

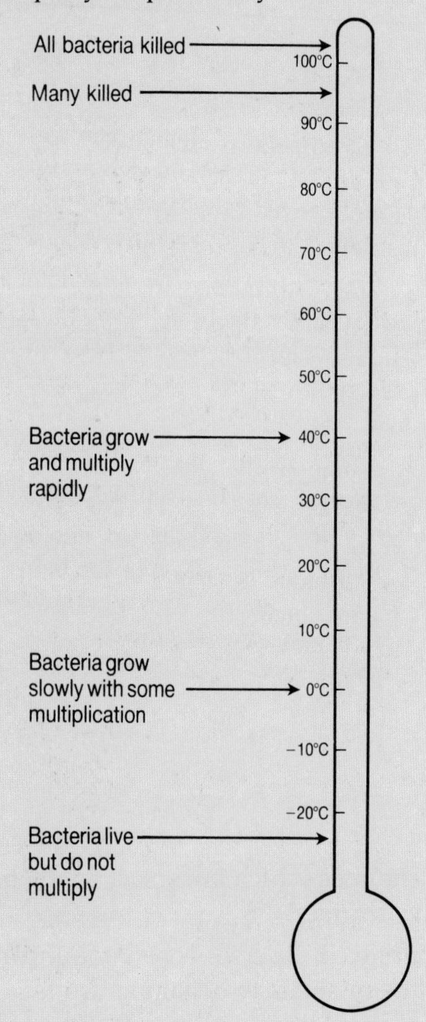

2 Why should you:

a) never use someone else's comb or towel?

b) always wash your hands after visiting the lavatory?

c) always wash your hands before handling food for other people?

d) never drink water from a pond or stream?

e) never bathe in the sea near a sewage pipe?

3 **a)** Write down the steps by which someone might become a heroin addict.

b) Write down ten reasons why drug-taking is harmful.

c) If drug pushers are caught they are put in prison. Do you think this is fair? Explain why.

4 Each cigarette shortens the life of a heavy smoker by five and a half minutes.
List as many reasons as you can, to explain why.

5 Here is some advice for people who want to give up smoking.

a) Give up smoking at the same time as a friend does.

b) Every day, put the money you would have spent on cigarettes in a money box.

c) Eat fruit or raw carrots, or chew gum whenever you feel the urge for a cigarette.

d) Tell everyone you know that you are giving up smoking.

Take each piece of advice in turn, and explain why it is worth trying.

6 Read the paragraph below, then answer the questions.
'AIDS is caused by a virus which can only live in blood, semen, and vaginal fluid. People can only catch the AIDS virus through sexual intercourse with an infected person, or by getting infected blood into their bloodstream.'

a) What is semen, and what is the vagina?

b) Why is it impossible to get AIDS by touching someone, or from cups, toilet seats, cutlery, clothes, door knobs or swimming pools?

c) Why does having many sexual partners increase the risk of catching the AIDS virus?

d) Why do drug addicts who inject themselves and share needles, increase their chances of catching the AIDS virus?

e) Why does the use of a condom reduce the risk of catching the AIDS virus?

Investigations

1 Build the cigarette-smoking maching shown in this diagram.

a) Using a suction pump, suck the smoke from at least five cigarettes through the apparatus. Now look at the glass wool and water, smell them. What do you notice?
b) What is the black stuff which is causing this smell?
c) List some of the chemicals which this black stuff contains, and the harm they do.
d) This is the stuff smokers suck into their lungs. Do you want it in yours?

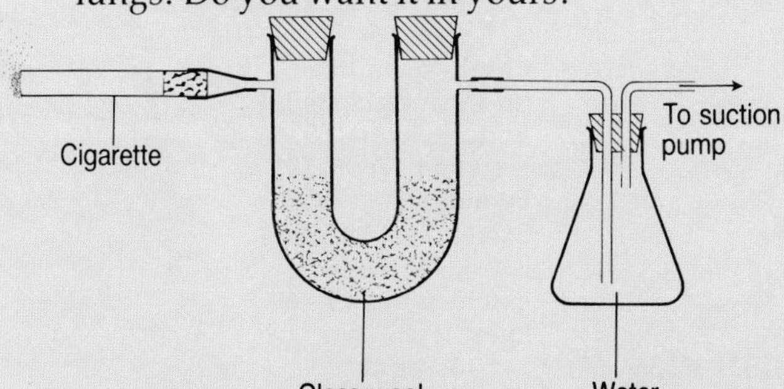

2 Investigate the smoking habit in your school. Circulate a questionnaire and use students' answers to make graphs and charts to illustrate your results. Make it clear than nobody will see anyone else's answers. Here are a few of the things you could try to find out.

a) How many smokers, non-smokers, and ex-smokers are there in your school? Calculate the percentage of each type.
b) What is the age of each smoker? Draw a histogram to show 'numbers of smokers' against 'age'.
c) Why do smokers smoke? Design a questionnaire with four or five reasons on it and ask students to choose the one that's true for them. Write a report on your findings. Do your answers vary with age and sex?
d) How many smokers would like to give up the habit? What percentage is this of the total number of smokers?
e) How many smokers are *trying* to give up? What percentage is this of the total number of smokers?

These are some ideas to investigate. You will think of many more. It's a good idea to conduct an investigation like this in a group. Think about **what** you want to find out, **how** you will find out, and **how** you present your findings.

3 Hold a smoking opinion poll. Ask as many people as possible to answer **yes** or **no** to questions like these, and use the answers to draw histograms.

a) Should smoking be allowed in: libraries, shops, cinemas, public houses, hairdressers?
b) Should cigarette advertisements be banned altogether?

4 What do people know about alcohol? Try this 'true or false' test on as many people as possible, then give them the correct answers. Draw histograms to show the number of correct and incorrect answers.

a) Drinking beer is less harmful than drinking wine or spirits. (False – all alcoholic .drinks can be as harmful as each other.)
b If a man and a woman drink the same amount of alcohol, the woman will be more affected than the man. (True – on average it takes less alcohol to damage a woman's health than a man's.)
c) Alcohol is a stimulant. (False – alcohol dulls the brain.)
d) Alcohol warms you up. (False – alcohol enlarges blood vessels in your skin so you may feel warmer when in fact your body is losing heat more quickly.)
e) Three pints of beer can put someone over the legal blood alcohol limit for driving. (True – an average-sized adult reaches the legal blood alcohol limit for driving after drinking about two-and-a-half pints of beer.)

5 Write a report on drug abuse in your area, or in the U.K. as a whole. You could get information for your report in the following ways:

a) Newspaper and magazine cuttings. Over a period of several months cut out any articles about drug abuse.
b) Interview people. Ask if you can interview the local police, a doctor, hospital staff, and District Health Authority staff, to obtain information about local drug problems. Some of these people may agree to visit your school and talk to your class.
c) Visit a local hospital (arrange it in advance) to ask what percentage of casualty patients became injured under the influence of drink or drugs.

Index

If more than one page number is given, you should look up the **bold** one first.

Acknowledgements

The publishers wish to thank the following for permission to reproduce transparencies:

Heather Angel: pp. 12 (centre), 142 (bottom centre), 145 (top); **Ardea**: pp. 141 (top), / **Ian Beames** 52 (right), / **Liz & Tony Bomford** 141 (bottom left), / **Ake Lindar** 141 (centre), / **P. Morris** 2 (centre), / **Wolfgang Wagner** 19 (right); **Aspect Picture Library**: pp. 152 (top), / **Derek Bayes** 142 (centre right); **Brian Beckett**: pp. 2 (top right), 3 (top right and bottom right), 8 (bottom left and bottom right), 9 (bottom left), 11 (bottom left), 12 (centre left), 13 (top right, centre left and bottom right), 16 (centre), 18 (all), 19 (top left, top centre and bottom left), 36 (centre left), 42 (both), 45 (left and centre), 47 (left and centre), 50, 51 (top left, top right and bottom right), 53 (both), 54 (top left, bottom left and bottom right), 55 (top centre, bottom left and bottom right), 56 (centre left, centre right, bottom left, bottom centre and bottom right), 57, 58 (all), 60 (all), 61 (left), 64 (bottom), 70 (bottom), 92 (centre left and bottom left), 96 (top right, centre and left), 97 (bottom), 109, 133, 135, 137, 142 (top), 143 (centre), 145 (left); **Biophoto Associates**: pp. 37 (top left), 55 (bottom centre), 92 (top right), 128; **Neil Bromhall**: p. 126; **Bruce Coleman Ltd**: p. 16 (bottom); **Colorsport**: pp. 3 (top left), 68, 71; **Gene Cox**: p. 35; **Daily Telegraph Colour Library/P. Beney**: p. 136; **Fauna & Flora Preservation Society**: p. 145 (right); **Sally & Richard Greenhill**: pp. 28 (right), 92 (top left), 122 (both), 130, 141 (bottom right), 143 (centre right); **Holt Studios Ltd**: pp. 36 (bottom left & bottom right), 54 (top centre); **Chris Honeywell**: pp. 33 (left), 54 (top right), 61 (right), 70 (top), 72, 79 (top), 81, 82 (top), 84 (left), 86 (left), 88 (top), 93 (all), 96 (right), 102, 105 (centre), 112, 129 (all), 151; **Imperial Chemical Industries**: p. 104 (top left); **International Coffee Organization**: p. 131 (bottom); **John Radcliffe Hospital, Medical Illustration Dept.**: p. 105 (top); **Rob Judges**: pp. 31 (both), 56 (top right); **Andrew Lawson**: p. 115 (both); **Network**: / **Mike Abrahams** p. 153 (bottom), / **Mike Goldwater** 142 (bottom right) / **H. Salvadori** p. 153 (top), **Oxford Scientific Films Ltd**: / **Doug Allan** p. 16 (top right): / **Kathie Atkinson** p. 17 (bottom right); / **G.I. Bernard** pp. 2 (bottom right), 10 (top centre and bottom left), 11 (bottom centre and bottom right), 13 (top left and centre right), 14 (bottom right), 15 (top right, centre left and bottom left), 16 (left), 28 (left), 37 (bottom left and bottom right), 43, 47 (right), 54 (bottom centre), 55 (top left), 65 (right), 67 (both), 113 (left), 120 (bottom right); / **David & Sue Cayless** p. 144 (bottom right); **J.A.L. Cooke** pp. 3 (bottom left), 4 (both), 9 (top left), 10 (top and bottom centre), 11 (top centre and top right), 13 (bottom left), 65 (left), 144 (left); / **Rob Cousins** p. 131 (top); / **Stephen Dalton** pp. 2 (top left), 12 (top right), 14 (top left); / **Michael Fogden**: p. 15 (bottom right); / **Phillip Goddard** p. 144 (top); / **Christian Hvidt/Foci** p. 44; / **Manfred Kage** p. 104 (top right); / **David Kerr** p. 45 (right); / **Caroline Kroeger/Animals Animals** p. 87; **Z. Lesczyski/Animals Animals** p. 113 (right); / **G.A. Maclean** p. 144 (bottom centre); / **Mantis Wildlife Films** p. 127 (right); / **Sean Morris** p. 120 (bottom left); / **Peter O'Toole** p. 108; / **Richard Packwood** pp. 8 (bottom centre), 17 (bottom left), 55 (top right); / **Peter Parks** pp. 8 (top left and top right), 9 (inset), 10 (top right), 12 (centre right), 14 (bottom left), 39 (all), 64 (top), 79 (bottom), 120 (top left); / **Avril Ramage** p. 15 (top left); / **Michael Silver** p. 17 (top); / **David Thompson** pp. 3 (bottom centre), 11 (top left), 16 (centre right), 51 (bottom left), 121, 139; / **Gerald Thompson** pp. 9 (bottom), 14 (bottom centre), 19 (bottom centre); / **Maurice Tibbles** p. 17 (centre left); / **Barry Walker** p. 16 (centre); / **Barrie Watts** p. 9 (top right); / **Nick Woods** p. 12 (bottom left); **Oxford University Press**: pp. 30, 32, 33 (right), 86 (bottom right), 98; **Picturepoint Ltd**: p. 143 (top); **Poly**: p. 36 (top and centre right); **Reflex Pictures/Philip Gordon**: p. 152 (bottom); **Science Photo Library**: pp. 77 (both), 97 (top); / **Michael Abbey** p. 8 (top centre), / **Dr Tony Brain** p. 22 (bottom left); / **Dr Arnold Brody** p. 83 (top); / **Dr Jeremy Burgess** pp. 23 (right), 52 (left); / **Dr Colin Chumbley** p. 116 (bottom); / **CNRI** pp. 10 (bottom right), 22 (bottom centre), 28 (top), 82 (bottom); 83 (bottom), 86 (top); / **CNRI/Professor Montagnier, Institut Pasteur** p. 148 (top); / **Eric Grave** pp. 22 (top), 120 (top right); / **Richard Hutchings** p. 37 (top right); / **Manfred Kage** pp. 22 (bottom right), 116 (top); / **Hank Morgan** p. 88 (bottom); / **Larry Mulvehill** p. 80; / **Petit Format/Nestle** p. 127 (left and centre); / **David Scharf** p. 124; / **Dr Gary Settles** p. 148 (bottom); / **St. Stephens's Hospital, Department of Medical Photography** p. 149; / **James Stevenson** pp. 84 (right), 150 (both); / **R.B. Taylor** p. 23 (left); **Supersport Photographs/Eillen Langley** p. 92 (top centre); **Patricia Taylor** p. 142 (bottom left).

Diagrams by: **Brian Beckett**

Additional Illustrations by: **David Holmes, Peter Joyce, Frank Kennard, Ed McLachlan, R.D.H. Artists**.

UNLOCKING EVIDENCE

2nd edition

Dr Charanjit Singh Landa
and
Mohamed Ramjohn

UNLOCKING THE LAW

LONDON AND NEW YORK

Second edition published 2013
by Routledge

2 Park Square, Milton Park, Abingdon, Oxon OX14 4RN
Simultaneously published in the USA and Canada
by Routledge
711 Third Avenue, New York, NY 10017

Routledge is an imprint of the Taylor & Francis Group, an informa business

First edition published by Hodder Education 2009

British Library Cataloguing in Publication Data
A catalogue record for this book is available from the British Library

Library of Congress Cataloging in Publication Data has been requested.
A catalog record for this book has been requested

ISBN: 978-1-4441-7103-7 (pbk)
ISBN: 978-0-203-78071-8 (ebk)

Typeset in Palatino
by RefineCatch Limited, Bungay, Suffolk

Printed by Bell & Bain Ltd., Glasgow